T0233306

Cambridge Elements ☰

Elements in Corporate Governance
edited by
Thomas Clarke
UTS Business School, University of Technology, Sydney

TRUST, ACCOUNTABILITY AND PURPOSE

The Regulation of Corporate Governance

Justin O'Brien

*The Trust Project
The University of Sydney*

CAMBRIDGE
UNIVERSITY PRESS

CAMBRIDGE
UNIVERSITY PRESS

University Printing House, Cambridge CB2 8BS, United Kingdom

One Liberty Plaza, 20th Floor, New York, NY 10006, USA

477 Williamstown Road, Port Melbourne, VIC 3207, Australia

314–321, 3rd Floor, Plot 3, Splendor Forum, Jasola District Centre,
New Delhi – 110025, India

79 Anson Road, #06–04/06, Singapore 079906

Cambridge University Press is part of the University of Cambridge.

It furthers the University's mission by disseminating knowledge in the pursuit of education, learning, and research at the highest international levels of excellence.

www.cambridge.org
Information on this title: www.cambridge.org/9781108748506
DOI: 10.1017/9781108781138

First published 2019

A catalogue record for this publication is available from the British Library.

ISBN 978-1-108-74850-6 Paperback
ISSN 2515-7175 (online)
ISSN 2515-7167 (print)

Cambridge University Press has no responsibility for the persistence or accuracy of URLs for external or third-party internet websites referred to in this publication and does not guarantee that any content on such websites is, or will remain, accurate or appropriate.

Trust, Accountability and Purpose

The Regulation of Corporate Governance

Elements in Corporate Governance

DOI: 10.1017/9781108781138
First published online: August 2019

Justin O'Brien
The Trust Project and The University of Sydney
Author for correspondence: Justin O'Brien Justin.obrien@thetrustproject.com

Abstract: The collapse of trust can be found across all of our institutions but most of all in finance. This Element seeks to answer an existential question: how to rebuild trust in distrusting times? Integrity, responsibility and accountability must be embedded into corporate mission statements, values and codes of conduct through organisational and regulatory design across five interlocking themes – legal, regulatory, managerial, ethical and social. What is required is substantive rather than technical compliance; warranted rather than stated commitment to high ethical standards; effective deterrence strategies; enhanced accountability; and a shared commitment to risk within negotiated, binding and enforceable parameters.

Keywords: trust, accountability, corporate governance, risk, integrity, social licence to operate

ISBNs: 9781108748506 (PB), 9781108781138 (OC)
ISSNs: 2515-7175 (online), 2515-7167 (print)

Contents

1 Introduction: How to Rebuild Trust in Distrusting Times

Growing up in a divided society, the borderlands of Northern Ireland, I acquired early the capacity to navigate the physical and invisible boundaries within and between people, place and community. It was training that served me well as an academic and as a television documentary maker. Both professions centre on stories. They require acute sensitivity to one's surroundings. Much of my work is informed by close ethnographic analysis of specific communities of practice and their perception of the application of rules, principles and social norms. It was how we survived and continue to do so.

I have explored this dynamic in settings as diverse as paramilitary organisations and New York trading floors. It has been critical to my work on the trilogy leading from Enron to the global financial crisis (GFC) – *Wall Street on Trial* (O'Brien 2003), *Redesigning Financial Regulation: The Politics of Enforcement* (O'Brien 2007) and *Engineering a Financial Bloodbath* (O'Brien 2009). It also informed a book on the intellectual history and limitations of the disclosure paradigm as a mechanism to ensure accountability and probity of publicly listed corporations, *The Triumph, Tragedy and Lost Legacy of James M. Landis* (O'Brien 2014a). This new work sketches how the corporation has impacted all of our lives at a global level. This Element is about learning to trust, the trade-off between vulnerability and hope and the danger of inaction.

Although I have been based in Australia since 2006, Ireland will always hold a special place for me. Its story permeates this work, not least because the financial dislocation there has reawakened dormant questions of the meaning of trust and the ongoing power of creative ambiguity to weaken it. It is indicative for me that when I return to Ireland, it is to an island eleven kilometres off the west coast, literally and metaphorically. Distance can provide perspective and the elemental power of the North Atlantic a reminder of the fragility of human existence. For those brought up in the British Isles, the daily 'Shipping Forecast' on *BBC Radio 4* is more than a navigational aid. It is an aural cultural icon that takes one on a journey around the divided islands, simultaneously providing a sense of space and serving as a beacon. Now more than ever, however, the danger lies within our divided borders not beyond them. Trust, essential to cooperation and development, is breaking down (Botsman 2017). With it an existential crisis has emerged one can no longer ignore (*Edelman Trust Barometer 2012–2019*). The history of that process and the lack of regard for the unintended consequences are, therefore, among the major research themes of our time: how to rebuild trust in distrusting times.

The collapse of trust can be found across all of our institutions but most of all in finance, a field I have made my professional focus for the past two decades. The wilful neglect of duty and responsibility and how this interacts with narrowly defined rights have led us to this moment. There is nothing inherently irrational in voting for Donald Trump in the United States or in the British people (by the narrowest of margins) placing membership in the European Union on a veritable burning of a bonfire of the vanities. Both sent powerful signals to governing systems that had lost touch with the primary drivers of a fulfilled life: the need for security, trust, purpose, confidence and hope.

Finance is critical to each. It has failed on all counts, with spurious promises to reform and expressions of regret by senior business leaders served up as mere platitudes (Agius 2010). At its core, the GFC and accompanying trust deficit were and remain moral failures. Technical measures alone cannot fix normative problems. Impotence, imprudence and irresponsibility combined to undermine the legitimacy of an entire world order, along with the erroneous theoretical assumptions on which it was based (Ulrich 2008). Now more than ever, there is a profound social, political and corporate imperative to rebuild – if not reimagine – the normative foundations of corporate governance and its regulation. It is essential to deliberate, to ascertain what are the duties and responsibilities as well as the rights of the corporation. Somewhat surprisingly, the quest leads us once again to the Antipodes.

Australia avoided the worst ravages of the initial GFC, more by luck than design. As I traversed the globe from 2007 onwards, the myopia encountered on my return to the so-called lucky country was staggering (O'Brien 2016). Belatedly, a Royal Commission into Misconduct in the Banking, Superannuation and Financial Services Industry was established in 2017, under restricted terms of reference and timelines. Contrary to expectations, the restrictions worked to its advantage. The result is not only an elegant disposition of the ethical malaise. It is also an innovative design based on core normative principles, which, if grasped, have the potential to be transformative, not only in Australia but also across the globe (Royal Commission, *Final Report*, 2019, pp. 8–9).

There are many similarities between building a personal and an institutional relationship, as the late Seamus Heaney (1998, p. 13) reminds us:

> Masons, when they start upon a building,
> Are careful to test out the scaffolding;
> Make sure that planks won't slip at busy points,
> Secure all ladders, tighten bolted joints.
> And yet, all this comes down when the job's done,
> Showing off walls of sure and solid stone.

The poetic metaphor is apposite. Commissioner Kenneth Hayne has put in place intellectual scaffolding that can be removed without compromising the integrity of the structure itself. It is a challenge to corporate, regulatory and political communities to put the interests of 'clients' (as citizens and fellow human beings) first. Following the emotional and social carnage of the GFC, it is also worth turning to another of Ireland's winners of the Nobel Prize for Literature, W. B. Yeats and his *The Second Coming* for further inspiration. In it, he warned of the danger of blood sacrifice, the attraction of believing in those prepared to die for a cause, the certainty of belief, the ease with which spurious solutions to manipulated crises can gain superior traction to reasoned argument. While he was referencing war, in uncertain economic times, Yeats (2000, p. 64) still has the capacity to chill:

> Things fall apart; the centre cannot hold;
> Mere anarchy is loosed upon the world
> The blood-dimmed tide is loosed and, everywhere
> The ceremony of innocence is drowned;
> The best lack all conviction, while the worst
> Are full of passionate intensity.

Each poet navigated the violence of his time in a different way. For Heaney, the aim was to hark back to less troubled times. He venerated honest toil where the digging was for crops not graveyards to fill with the patriot dead. A pen not the Armalite was to be the instrument of most utility. Yeats, on the other hand, was avowedly political. The Easter Rising of 1916 and subsequent execution of its leaders were immortalised in the oft-quoted lines that 'all changed, changed utterly; a terrible beauty is born' (Yeats 2000, p. 60). This flirtation with violence dissipated as the world slipped towards chaos, the result of rage, resentment and the return of the politics of the strongman. As with the 1930s, there is a profound sense of fin-de-siècle. For Yeats (2000, p. 92), in the end what mattered most in making what he called *The Choice* was a solid emotional core:

> When all that story's finished, what's the news?
> In luck or out, the toil has left its mark:
> That old perplexity an empty purse
> Or the day's vanity, the night's remorse.

Our capacity to make those choices, however, is always informed by trust or its absence. One can be cognisant of the past without being imprisoned by it. It most certainly does not have to predetermine our future. It is not in our interest that it does. In an age in which opinion trumps expertise, all that is left are appeals to common sense and vague replies to questions of substance. This is fertile ground for the populists of whatever ideological hue, or none, to furrow. In creating foundations of cynicism, they succeed only in building a Potemkin façade. This Element seeks to address those architectural defects. It provides the evidential basis on which the normative claims for a reconfiguration of corporate duty, purpose and obligation are made.

2 The Existential Crisis Facing the Liberal Order

Trust can be defined as having firm belief in the reliability, truth or ability of someone or something not to abuse the vulnerability we put at risk in extending confidence (O'Neill 2002, 2016). The foundation of trust in the Western liberal state, however, is disintegrating (Fukuyama 1995). So too is warranted faith in two critical components of institutional trustworthiness: competence and honesty (Hosking 2014, p. 108). The disdain seeps across all sectors – from business to government, charitable bodies to media (*Edelman Trust Barometer 2012–2019*). Fake news, which that master manipulator Orson Welles reminds us, is as old as the Garden of Eden, is emblematic of an age informed by the corruption of knowledge (Schwartz 2014, p. 51). From meddling in the US presidential election to Brexit and beyond, the capacity to engage in rigorous policy formulation, implementation and evaluation has been compromised by communication channels that amplify and distort (Pomerantsev 2017; Kakutani 2018; Kavanagh and Rich 2018).

We are, as Salvoj Žižek (2017) reminds us, living in 'an age of hopelessness'. It is a time and place where

> true courage is not to imagine an alternative, but to accept the consequences of the fact that there is no clearly discernible alternative: the dream of an alternative is a sign of theoretical cowardice, functioning as a fetish that prevents us from thinking through the deadlock of our predicament. In short, true courage is to admit that the light at the end of the tunnel is probably the headlight of another train approaching us from the opposite direction. (Žižek 2017, pp. xi–xii)

These are not just the fatalistic pre-occupations of the radical left, which sees in populism an opportunity to re-energise a lost socialist dream (Mouffe 2018). The fears also reflect the pragmatism of the international foreign affairs establishment, the most important journal of which offers dystopian futures ranging from realism, liberalism, tribalism; a return to Marxism; an over-reliance on technology; to the socio-economic and geo-political implications of global warming destroying the planet (*Foreign Affairs* 2018).

The *Edelman Trust Barometer* (2012–2019) traces the uncertainty to the 2007–2008 GFC and its subsequent management. There is merit in the case

(Muller 2016; Nicholls 2017). The parsimony, however, ignores broader questions about accountability of corporations and the effectiveness of oversight, which necessarily must be evaluated against a defined purpose (O'Neill 2002; Ulrich 2008). These concerns have long informed debate on the ethical challenges of the corporation (Berle and Means 1932; Mason 1960; Friedman 1970; Freeman 1984, 2010; Smith 2005, 2006; Clarke, O'Brien and O'Kelley 2019). Disputes over legal form and discrepancies between legal form and purpose have been ongoing for more than a century (Orts 1993; Halperin 2011). They remain unresolved in both 'law on the books' and 'law in action' if not in the emaciated conception of 'economic man' or the corporation as a mere nexus of contracts (Coase 1937). This is accompanied by a misunderstanding of the philosophical underpinning of Adam Smith's emphasis on 'empathy' rather than 'self-interest' as the driver of social progress (Rothschild 1994; Norman 2018).

By illuminating how the economically rational is in itself a political construct (e.g. Fligstein and Dauter 2007), an admission belatedly accepted by a former chair of the Federal Reserve (Greenspan 2008), this Element deconstructs dangerous myths propagated for narrow political and educational status rather than social advance (Ulrich 2008). Their accretion gave the illusion of substance. Instead, the model buckled under the weight of its own internal contradictions (Bell 1996; Judis 2016; Sunstein 2018). In so doing, it revealed the idealistic yet hidden ideologically driven assumptions of the law and economics and institutional economics traditions. Each maintained a focus on the transactional over the relational through an emaciated reading of political theory (Frank 2001; Frazer 2010; Kuttner 2018). Each also confused economic assertion with empirical fact on the legal parameters in which the corporation operates and its underlying purpose and control (Berle 1931; Dodd 1932; Weiner 1964; Stone 1981; Orts 1993; Stout 2007).

The status of the corporation and what are or ought to be its rights, duties and responsibilities within this legal framework have waxed and waned since its initial status as a grant of royal prerogative. Limited liability, facilitated by political dispensation, gave both impetus and extended power, which have intensified with the rise of globalism and the information technology that facilitates it. It is deeply ironic that virtual communities have been shaped by Silicon values, imbued with a libertarian ethos that combines naiveté with calculation (Cohen 2017, p. 4). A misreading of Hayek's (1943) emphasis on market solutions and a disregard for how disconnecting the state from the market is a political choice (Polanyi 1944) lead to a facile veneration of 'creative destruction' (Schumpeter 1943). What is lost is the prophetic warning that the stock market is a poor substitute for the Holy Grail (Schumpeter 1943, p. 129), and that capitalism's main threat come from within. This Element ties together these disparate strands to weave a narrative of practical and theoretical value, with applicability across the disciplines.

Advances in corporate governance theory, presupposing a 'communitarian' turn (Millon 1995; Mitchell 1995) or a 'team-production model' (Blair and Stout 1999), while laudable, have failed to situate the corporation as an institution within a wider context of social responsibility in conceptual (Miller 2017), philosophical (Parfit 2011) or practical terms. Ulrich (2008, p. 2) asserts it is 'not markets but citizens [who] finally deserve to be free in modern society. The market economy must, therefore, be civilized in a literal sense.' Indeed, within the management if not economics literature, this precise failing was identified nearly six decades ago (Mason 1960, p. 19). Each move to reform, however, was occasioned not by acceptance of responsibility. Instead, responses to crises and scandal were reactive and minimal (Rossouw and van Vurren 2003). What then, as Lenin (1902), famously exhorted, is to be done?

In a previous era, the Italian Marxist theorist Antonio Gramsci (1971, p. 35) advocated policies designed to raise the consciousness of the nation. This was the *raison d'être* of the political party, deemed by Gramsci, following Machiavelli, to be 'The Modern Prince'. Today, as Sheldon Wolin (2016, p. 616) reminds us, the task is not a retreat to the failed politics of nationalism but a regeneration of what ought to be the foundational requirements of public socio-economic security (see also Piketty 2017, 2018).

Sustainability requires an acknowledgement that citizenship is not a passive administrative right. It brings with it a series of duties and responsibilities, of which electoral participation in a human sense and influence through corporate spending are only component parts. How does one make the necessary deliberative governance legitimate and effective? At stake is an unavoidable and indeterminate conflict within and between varieties of capitalism. Critical to this process is a mapping and navigation of the turbulent relationship between the corporation and the multiple societies it impacts. Before we can address a pervasive sense of hopelessness, we have to first understand it.

2.1 The Corporation as Problem and Solution

Nowhere is the trust deficit more apparent than in the influence of the corporation both on individual lives and in polities beyond market economies, not least because of the rise of state capital on capital markets themselves, where entities directly connected to governments are increasingly active players. 'Capitalism has transformed itself, from a system of activities analysable through economic categories to one that has adapted political characteristics and the qualities of a new constitutional blend devoid of democratic substance,' noted Wolin (2016, pp. 587–90). The light coming down the tunnel has proved to be that of an approaching train. The ticket on the journey to globalism is paid for, but its passengers are increasingly subservient to forces beyond their control. As Wolin (2016, p. 588) wrote, 'the cooperation of corporate power is now a vital element of domestic, foreign, and military policies. Competition and rivalry occur less between state and corporations and more between corporations vying for influence over the state or subsidies from it.' Individual self-interest is presented in positive terms as evidence of an entrepreneurial spirit which is both rational and inevitable (Downs 1957, p. 295). It is a beguiling if not bewitching story. It is one that is given added plausibility by technological innovation, which its libertarian founders proclaim would be threatened by any regulatory restriction on moral, cultural or political grounds. The corporation, across multiple sectors, most notably finance and technology, has become both the guardian of economic growth and the locus of democratic vulnerability. If ever it were capable of acting as sentinel, the corporation has never before been so unsuited to the task.

In August 2018, Apple, a once-struggling computer manufacturer, rose from the ashes to make history as the first corporation to reach a market capitalisation of more than 1 trillion US dollars. In all of the plaudits, less remarked upon is the source of that innovation. For Apple, this lies as much in its tax strategy and ability to skirt around regulatory rules designed for a previous epoch as its technological prowess and marketing genius (Senate 2013). Meanwhile, manifestoes to create global virtual communities, encapsulated by protestations from Facebook that 'fake news is not our friend' ring as hollow as the finance industry's apologies. In the ten years since the GFC, technology has indeed changed our world. This has not necessarily been for the better. The recklessness of the financial services sector brought the global economy to a standstill. Maladroit oversight offered in advance, during and after the crisis did much to calcify cynicism in the developed world about regulatory competence. A hardening agent was provided by the alleged democratisation of the news cycle offered by social media.

There is an unnerving similarity in the responses of a financially (if not morally) chastened global finance industry and its tech counterparts to transparency and accountability deficits. Both have proved impervious to effective control. They remain so. Expressing regret is not the same as accepting responsibility (O'Brien 2014b). Disruption, like innovation, is not a value-neutral noun or concept (Cohen 2017). Everything has a price. Far from complying with generally accepted principles of best-practice corporate governance, these highly mobile, asset-light corporations are adroit at aggressive tax minimisation strategies. These sail uncomfortably close to tax evasion. Illegal state aid facilitates a race to the bottom in the design and application of foreign direct investment strategies, even for those, such as Ireland, that operate under common market rules (European Commission 2017).

The competitive advantage of the multinationals is enhanced by deft exploitation of gaps provided by the non-applicability of traditional systems of oversight. These gaps include lack of responsibility for the curation of hosted content, the ideational privileging of innovation and disruption over stability and the suggestion that curbs on operational independence violate libertarian principles. Cavalier approaches inform methods of data protection, and technical gaming of regulatory rules designed for an industrial economy destined to history because of the rise of artificial intelligence and its impact on the future of work itself display a confidence that borders on hubris (Cohen 2017).

In seeking to restore warranted confidence in corporations and their governance, this Element explores how wealth generation can operate within sustainable frameworks, keeping in creative tension the societally beneficial goals of liberty and equality (Rousseau 2005). Maintenance of social order and progress requires active balancing of both. If equality of opportunity is compromised by a privileging of individual freedom, the social contract disintegrates. This can lead, inexorably, to a corresponding increase in resentment, which is playing out across the developed world. Distrust damages social cohesion and quickens a downward spiral, as Arendt (2017, pp. 4–5) recounts in the seminal *The Origins of Totalitarianism* (see also Adorno 1950).

Arresting this malaise requires a fundamental change in thinking. We all know (or can find legal representation to litigate) our corporate rights. The question of what constitutes duty and responsibility is much more vexed, not least because most corporate constitutions are written in such a way that they include provisions that facilitate the conduct of anything lawful. What is lawful, in turn, depends on whether the definition of lawful conduct is read narrowly or purposively (Braithwaite 2013). Given that so few cases taken by regulatory or prosecutorial agencies are litigated to a judicial conclusion, there is a profound absence of precedent. This has knock-on effects across common law systems (O'Brien 2007; Rakoff 2014, 2019). Administrative agencies have a duty to act as model litigants, meaning they have to have a reasonable belief in the veracity of their case (O'Brien 2013). Without a precedential guide, these agencies are susceptible to accusations of overreach. It is the oldest playbook in contemporary democracy and has compromised the authority of the administrative process from the beginning (Landis 1938, 1960; O'Brien 2014a).

2.2 The Rise of Precariousness

In today's globalised world, no physical wall can provide a protective shield. The challenge is, therefore, as much virtual as physical. As the dislocation caused by technological disruption extends and expands, middle-class aspiration becomes increasingly, if paradoxically, unattainable to achieve or to hold. We are witnessing a global rise in 'precariousness' (Standing 2011). Anaemic economic growth and the erosion of economic certainty and job security are advanced within and across the 'gig' economy. These are often presented as positive developments. They are anything but. The author of the term 'precariat' now sees it as an affliction increasingly impacting the professions, including law. In a 2017 interview, Guy Standing argued,

> [T]he division of labour has changed. For example, the legal profession has silks at the top, then salariat lawyers and a huge growth of paralegals with basic training, but without a career path through the profession, because there are ceilings. It's the same in the medical and teaching professions. The precariat is not just a reality, it's grown extremely fast in recent years. Unless something is done to improve security and redress the class-based inequalities arising, you're going to get a political monster. (Cited in Johnson 2017; see also Piketty 2018)

The investigation undertaken here to address these problems is explicitly comparative. It enables evaluation of corporate purpose and obligation across jurisdictional and disciplinary boundaries. It focuses on the individuals, functions, processes and purposes through which the corporation is directed and controlled. In so doing, it designs and tests how more granular principles of corporate governance can incorporate essential conceptions of fairness and justice. In technical terms, this is a design blueprint of 'integrative economic ethics' in action: 'an ethically rational [re-]orientation in politico-economic thinking without abandoning reflection in the implicit normativity of "given" economic conditions' (Ulrich 2008, pp. 3–4). This, in turn, displaces transactional approaches that equate efficiency with fairness.

Recalibrating the duties and responsibilities of the twenty-first-century corporation requires a verifiable process of outlining a rationale, ongoing deliberation, implementation, monitoring, assessment and subsequent evaluation. Delivered through evolutionary methods, its success would herald a revolutionary outcome: the reshaping of corporations and markets to deliver inclusive economic growth, leading to better lives in line with the integrative economic ethics framework suggested by Ulrich (2008). This agenda is not one dreamt up in an ivory tower. It is an agenda recognised as essential by the Organisation for Economic Cooperation and Development (OECD 2018). It is, moreover, an agenda capable of application in non-OECD jurisdictions.

Effective corporate governance and capital market regulation are as much imperatives in Moscow as Frankfurt, London as Beijing, New York as Johannesburg and Brussels as Riyadh. Global in (real) life but national in death, the corporation poses wicked questions to ongoing domestic political legitimacy, irrespective of domain. Whether, where, how and indeed how much tax a corporation pays have profound implications. If taxes are not collected, the trickle-down logic of profit as sole source of purpose evaporates (Friedman 1970). The rise of state capital and expansion of state-owned enterprises and sovereign wealth funds (SWF) as key actors in capital markets further complicate matters. The listing of red chip corporations in Hong Kong, London and New York is as problematic and as politically charged as the shareholder activism of the wider SWF sector (or its private equity equivalents). The governance structures of the Saudi Arabian oil giant Aramco, poised to be the largest single initial public offering in history, plays second fiddle to the fees associated with the transaction. Whether shareholders will have any real power to influence its decision-making is, however, as questionable as control of the technology giants, all of which use differential voting shares to ring-fence and protect the founding leadership. For all the hype surrounding disruption and innovation, we have by default allowed the creation of oligopolies on which we increasingly depend, both physically and mentally. And it is the failure to price this cost that is driving the trust and its discontents narrative.

2.3 The Building Blocks of Normative Value

I propose five core criteria essential to building a sustainable theory of the corporation that holds rights, duties and responsibilities in creative and productive tension. These are closer in roots to Adam Smith's *Theory of Moral Sentiments* (2006) than *The Wealth of Nations* (2005). The latter sets out the minimal conditions required for a functioning market; the former establishes the normative criteria for an examined and fulfilled life lived with integrity (Aristotle 2004). First, I demonstrate the need to rebuild public trust and confidence in corporations and those individuals and organisations responsible for their internal and external oversight. The very fact that much of what happened in the GFC and continued even after extensive bailouts was legal if irresponsible and unethical did much to undermine the foundations of trust within and between political, regulatory and corporate spheres. Notwithstanding the necessity at the time of providing guarantees to the banking sector, which had become too big to manage, fail, jail or control, the unanswered questions are whether the compromise brokered was fair, reasonable or sustainable. As Arendt (2017 p. 5) famously put it, 'neither oppression nor exploitation as such is ever the main cause for resentment; wealth without visible function is much more intolerable because nobody can understand why it should be tolerated.' Without reintegrating law and morality within definable and measurable frameworks, we risk making ever more resilient a system that is palpably unfair and unsustainable. A shift in rhetoric in and of itself is insufficient (G20 2018).

Second, as we move from crisis to strategic management, there is an urgent need to embed integrity, responsibility and what the philosopher Onora O'Neill (2002) terms 'intelligent accountability' (individual and corporate) into corporate mission statements, values and codes of conduct. The aim should be to facilitate ongoing learning rather than punitive reactive assessment (Ellison 2012, p. 32). This can be achieved through improved organisational and regulatory design across five interlocking themes – *legal, regulatory, managerial, ethical* and *social*. In summary, what is required is substantive rather than technical compliance; warranted rather than stated commitment to high ethical standards; effective deterrence strategies; enhanced accountability; and a shared commitment to the facilitation of risk within negotiated, binding and enforceable parameters. The 'non-calculative social contract' on which the modern governance system is based (Williamson 2000, p. 597) has been rendered void by neglect and entitlement in equal measure. To rectify this, the interlocking aims of *compliance, ethics, deterrence, accountability* and *risk* must be measured against *mandate, agency* (or discretion) and *process*, giving a holistic five-by-three matrix of investigation and evaluation of equal applicability to regulated and regulator.

The lost shared commitment of the non-calculative social contract is an essential component of the third aim, which builds on an essential if underutilised resource, the work of business ethicist Peter Ulrich (2008) and his theory of 'integrative economic ethics'. This is integrated with the abstract philosophical reasoning of Derek Parfit (2011, p. 321), who argues that an act is moral if it is 'universally willable, socially optimific and not unreasonably objected to'. Taken together, they provide a route map for practical tangible outcomes capable of verification.

The corporation operates within an overarching regulatory and social architecture. In many ways, this architecture can be understood as a complex adaptive ecosystem. This has both physical and cultural or ideational dimensions. How the corporation operates is determined by the interaction among potentially conflicting legal rules, principles and social norms. This necessitates evaluation of the symbiotic relationship between conduct and culture (within corporations, regulatory agencies and political systems). How social norms are created, embedded or (where necessary) diluted or evaded is a critical if under-explored area of regulatory theory and practice (Bicchieri 2017). Excavating the sources of those norms, their dynamics and their legitimacy is critical in under-standing the rules of the game (Ulrich 2008).

The fourth aim is to identify, rank and find solutions to obstacles to effective performance. These include structural factors, bureaucratic processes, manage-rial imperatives and remuneration procedures. The emphasis on short-termism is an affliction that extends far beyond either big tech or the financial services industry. In all aspects of life, our attention spans are shortening. We face multiple paradoxes. Never before has so much information been made avail-able; yet, quantity and quality are demonstrably in conflict. It is less a question of an absence of expertise; rather, gradual erosion of confidence in the veracity of information and the motives of those making the claims informs discourse (Nicholls 2017). This takes us to the fifth objective. What is to be done?

The title and substance of Vladimir Lenin's (1902) rousing polemic have been appropriated by the alternative-right movement. Lenin's self-stated objec-tive was to engineer the state's destruction from within. It is one thing to recognise structural defects to rectify them (or, in Lenin's case, to tear the structure down). It is quite another to weaken the one entity that can through legislation and its application weaken the deleterious hold the corporation has over societal preferences (*Four Corners* 2018). The same illogicality applies to the British government's response to withdrawal from the European Union. Taking back control has appeal as a campaign slogan. It is less effective in offering a strategic recalibration of the corporation and the state in which it is nested.

We cannot understand the erosion of particularised and generalised trust unless we understand the micro drivers of social inclusion and societal belong-ing (Yamigishi and Yamigishi 1994; Fukuyama 1995). These, in turn, are central to the relationships within and between corporations and the legitimacy of corporate activity. While these are manifested in all aspects of contemporary political and economic discourse, they have been most apparent by their absence in finance. To understand our modern trust deficit, and how we can respond to it, we must understand the relationship between this untrustworthi-ness and corporate law and governance. This cannot be done without attending to corporations and their role in deepening social fractures and entrenching economic inequalities. Furthermore, in an environment of eroding trust, we need to understand the extent to which, and how, law, regulation and technology can mend these fractures by deploying effective trust substitutes.

2.4 Managing the Trust Deficit

With a distinct legal personality, the corporation has the right to enter into contracts. It is protected at law by the principle of limited liability. This leaves it exposed only to the value of its assets. Unlike the shareholders who hold stakes in the company, the corporation itself can last in perpetuity, so long as its board of directors, its controlling mind, fulfils a fiduciary duty at law to act in the entity's long-term interests. These actions may conflict with the interests of individual shareholders, including those with substantial economic or voting interest in current or future strategy, or the lobbying interests of powerful interest groups. Over time, attempts have been made to address the separation of ownership and control (Berle and Means 1932), including the number and power of independent directors, those who hold a strategic oversight role but do not exercise executive power. In reality, neither they nor shareholders, better described as 'rentier of capital' (Ireland 2000), have either the access to information or the capacity to shift the strategic direction of the corporation (Weiner 1964; Bratton and Wachter 2008). That remains exclusively in the hands of management, the only restraining force to which it operates is its stewardship obligation (Dodd 1932), a position Berle himself was to concede as the only tenable one (Mason 1960).

Herein lies the problem. With directors having no overarching purpose or duties and responsibilities beyond operating within the law, one is left only with trust that they will fulfil their fiduciary duties. That faith can no longer be relied upon (Braithwaite 2013; O'Brien and Gilligan 2013). To rebuild warranted trust by the public, we need to hold corporations to account for deviance from core principles. To be effective, it must be internalised, not given lip service. Promises must matter (Mansbridge 1999). No theory of the corporation or the markets in which it operates remains sustainable without this ethical under-pinning. It must, however, go further than that. The corporation has a duty to uphold not just its own integrity but also that of the market. It has a responsibility to ensure that its own actions or inactions do not compromise that integrity.

This is what Ulrich (2008, p. 2) terms the 'civilizing context'. This cannot be ad hoc. It must be articulated, curated and evaluated. The precise contours of these duties and responsibilities will vary depending on cultural context. They must conform to universal standards, hence the importance of the normative framework offered by the Australian Royal Commission into Misconduct in the Banking, Superannuation and Financial Services Industry (*Final Report*, 2019, pp. 8–9):

- do not mislead or deceive;
- provide services which are fit for purpose;
- when acting for another, act in the interests of the other;
- obey the law;
- be fair; and
- deliver services with reasonable care and skill.

These are not unreasonable starting points in both theoretical and practical grounds. First, they unite categorical imperatives, utilitarian or consequential and virtue-based ethical schema as set out by Parfit (2011, p. 321). If shame and guilt are defined by departures from community expectations, what are they? How and through what forum are they to be judged? These are by no means simple questions. They cut to the core of how one defines responsibility, accountability and trust. They are essential to the potential utility of the Hayne report. Hayne's principles are a start, but only a start. What does acting fairly mean? When Hayne suggests that services must be fit for purpose, what purpose? What does 'reasonable' mean in the provision of services? When acting in the best interest, under what if any criteria? One can remain in technical compliance with the law but in breach of its moral framing. More specifically, this forces one to define the role played by culture in embedding, creating and sustaining general and specific social norms. It is a task that has increasingly exasperated scholars and critics throughout the twentieth and twenty-first centuries (Elliott 1948; Steiner 1971; Vargas Llosa 2015; Eagleton 2016). As we shall explore further, cultures can be repressive as well as enlightening. They can privilege elitism as well as denigrate individual self-expression. They can legitimate authoritarianism even within democratic frameworks. They are as much instruments of control and hegemony as bulwarks of civilisation. Critically, they impact the entire polity. Just as an individual influences culture, so too does a culture influence the individual and the institution he or she is affiliated with.

Second, however, given the systemic nature of the problem, there is no alternative. The scale of the task becomes clear when one considers three core interconnected threats and opportunities. First, implementing effective strategies to deal with the financial threats to national exchequers posed by base erosion and profit shifting (BEPS) lies at the heart of an unachieved agenda advanced by the OECD. A second acknowledged threat is the danger of (deliberate or inadvertent) corporate corruption of political processes. A third is deterioration in confidence in the governance of corporations themselves, particularly in the technology sector. Notwithstanding corporate advertising in support of counterculture West Coast cool, there is nothing socially engaging about tax strategies that blur the line between legitimate tax avoidance and illegal tax evasion (Senate 2013).

Progress within and between each priority is, therefore, examined and evaluated throughout this Element by an integrated set of qualitative and quantitative metrics. The insertion of normative goals (e.g. using corporations and markets to fund inclusive economic growth leading to better lives) allows for systematic examination of the purpose of the corporation and its relationship with society. The advance of populism represents not just anger and resentment. It also reflects the loss of legitimacy and authority (Muller 2016; Snyder 2018). We must integrate social, political, legal and economic dimensions. This integration reduces the trust deficit, aligns corporate purpose to changed societal expectation, complements legal obligation within a defined and measurable social licence to operate and contributes to more effective and accountable regulatory strategies capable of securing public support (O'Brien et al. 2015). It is time to deliberate.

3 Resilience as the Organising Framework for Reform: The Dangers of Metaphors in Financial Regulation

Purpose and vigilance are essential guarantees for the stability of any complex system. Responding to this crisis through resilience as both metaphor and organising framework is, however, problematic. Notwithstanding its increasing usage, *resilience* is not a neutral concept. Privileging resilience as an end in itself may prove counterproductive unless underpinned by a normative reset of the purpose of the corporation and the market and duties and responsibilities for vigilance each owes to society.

3.1 Inside the Beehive

The normative and practical value of resilience can be likened to the circumambulation of a beehive. It is here where service, industry, preservation and order integrate to achieve social harmony. In structure, purpose and functioning, the hive is the antithesis of uncertainty. It is a compelling metaphor in an age of anxiety and distraction (Mendelson 2016; Mathews Burwell 2018; Smee 2018). Unlike the hive, in the human social world, common purpose, duty and obligation are weak and weakening (Edelman 2012–2019). While the neo-liberal market consensus may have broken down (Greenspan 2008; PCBS 2013; Carney 2015), libertarian impulses that preservation of self-interest will safeguard other-regarding stability remain stubbornly resilient within both capital markets (O'Brien 2013, 2016; Awrey and Kershaw 2014; Moorhead and Hinchley 2015), and sectors beyond finance (Cohen 2017). Governance through self-interest alone, however, guarantees disloyalty and chaos.

For Adam Smith (2005), the invisible hand of the market was a metaphor to describe the minimal conditions necessary for the creation of a commercial market (not an industrial or post-industrial one, which was to become a reality long after his death) and certainly not a society that would lead to personal or social fulfilment (see Smith 2006). Likewise, the social contract model as envisaged by Rousseau (2005) was informed by fear about its fragility. Vigilance against self-interest was and remains necessary. Each philosopher provides essential contemporary warnings.

The contemporary liberal order now faces profound questioning of its capacity to resolve intractable problems. Most manifest on the streets of Paris in December 2018, de-synchronised disorder has fuelled an anti-establishmentarian movement that pits a mythical 'us' against an equally amorphous 'them' (Bremner 2018; Collier 2018; Deenan 2018; *Financial Times* 2018b). Life lived with meaning requires dignity (Aristotle 2004). Purpose requires more than transactional certainty. Equality of opportunity works as an essential counterbalance to liberty. Along with rights come duties and responsibilities. The precise form they take is an exercise of continual deliberative governance in which ethics must be integrated into corporate practice (Royal Commission *Final Report* 2019).

Paradoxically, it has been left to unelected, unaccountable policymakers to navigate a passage towards compromise and authenticity. It is an uphill task worthy of Sisyphus, the anti-hero of Greek myth destined to endlessly carry a boulder up a hill only to see it fall (Carney 2018). All that is left to break this cycle are (hollow) assurances to (undefined) audiences that (partial) technical measures can and will address structural and normative problems. The measures, how they are negotiated and then disseminated, display a tonal deafness to a global zeitgeist. In part, this is because of a generic disdain towards expertise among the wider populace (Nichols 2017). In part, it also reflects limitations in the mandates of both domestic regulators and international coordinating bodies. Power weakens the further one departs one's own territorial waters.

The Financial Stability Board (FSB), for example, can recommend. Transposition into domestic law even within the G20 itself remains out of its control. To complicate matters further, there is a world of difference between law on the books and law in action, a legal debate that has been raging for more than a century (Pound 1910; Halperin 2011). A further problem is the capacity of the transnational firm to transact around obligation by negotiating the content of soft law norms (Teubner 2002; Porter and Ronit 2006; Stone Sweet 2006; Picciotto 2011). In all cases, the strategies needed require the very thing that is absent: trust in the trustworthiness of the messenger. When claims are made that considerable effort is shoring up the resilience of any given system, just what does that mean? What measures are being used to demonstrate actual fealty? Is resilience or, more accurately, the definition of resilience used to defend the engineering integrity of the architectural blueprint offered by the G20 and by extension the FSB the answer? This section suggests not, or at least not in the way it is currently being used.

3.2 Resilience as Metaphor or Organising Concept?

'Resilience' is replacing 'sustainability' as both the means and the end of an increasingly cross-pollinated academic and policy discourse. On one level, this is unsurprising. Resilience suggests strength and adaptability. It infers more capacity to identify and neutralise threats in advance which will enhance and protect well-being. As such, it offers a beguiling trifecta: diagnostic tool, palliative treatment and ongoing sentinel protection, each of which spans temporal and spatial horizons (Brown and Westaway 2011). The more detailed the proposed approach, in theory if not in practice, the more confidence is engendered. These measures can include risk indicators that emphasise inter-connections between environmental, social and economic aspects to determine actual and perceived quality of life (Stiglitz et al. 2009; G20 2017a, 2017b; Stiglitz et al., 2018). If, and only if, credible in action, they can offer a baseline for reform.

This claim, for example, informs the latest G20 (2018) communiqué from Buenos Aires. Its leaders claimed the policy agenda had been refocused (if belatedly) towards fairness and sustainability, without actually defining either. Specifically, the G20 (2018, para. 25) maintained that an 'open and resilient financial system, grounded in agreed international standards, is crucial to support sustainable growth', while noting but downplaying the danger of 'fragmentation'. None of this is actually spelt out. To his credit, the outgoing chair of the FSB, Mark Carney (2018), was less sanguine about a world in transition. Carney (2018, p. 2) warned about the danger of complacency and the need for vigilance, in non-bank finance and asset management; operational risk, particularly from cyberattack; and the potential 'issues crypto assets pose around consumer and investor protection, money laundering and terrorist financing'. He also emphasised that the wholesale banking system itself was becoming more resilient, not least because of design of a

> sophisticated toolkit of measures which supervisors and firms can use to strengthen the governance frameworks of financial institutions, including by increasing the accountability of senior management for misconduct within their

firms. The toolkit complements other elements of the FSB's Misconduct Action Plan, including compensation recommendations that better align risk and reward and reforms to strengthen major financial benchmarks, as part of broader measures to restore public trust in the financial system. (Carney 2018, p. 4)

A toolkit, however, is only as useful as the skill of those deploying it with specific outcomes not outputs in mind. As Carney (2018, p. 6) concluded wistfully in his final report as chair of the FSB, '800 years of economic history teaches that as memories fade, complacency sets in, and backsliding begins. In financial stability, success is an orphan. The G20 and FSB bear heavy responsibilities to safeguard recent progress, address new risks, and seize new opportunities presented by the major transitions underway in the global economy and financial system.' One does not need advanced code-breaking skills to decipher the bluntness of the message.

In a paper released to coincide with the G20, the Group of Thirty (an influential group of former bankers, regulators and policymakers based on K Street in Washington, DC, the epicentre of lobbying influence) released its own warning. It claimed the finance industry and its regulators faced four unchanged interlinked problems (G30 2018, p. xi): first, a public perception that nothing has actually changed; second, the danger of fatigue setting in at corporate, regulatory and political levels; third, uncertainty about the impact of technology; and fourth, fear that changes already embedded will be progressively hollowed out. The risk is magnified if the tone at the top or from above, as the G30 puts it, does not or cannot break through the permafrost of middle management to change actual practice within myriad subcultures, each governed by its own set of social norms.

Recognition of the problem is not the same as designing workable, scalable solutions. The G30 delivers an implicit rebuke of the widespread adoption of 'the three lines of defence' approach to risk management – identification at the business unit level, second level performed through risk and compliance with internal and external audits providing third-level validation of the efficacy of controls to both the board and investing public (G30 2018, p. xii). If the risk is not identified at the first level, the second and third are inevitably compromised, resulting in a flawed prospectus on which to base appeals to trust (Arndorfer and Minto 2015). As should now be clear, the identified defects begin to erode confidence in the overarching G20 and FSB strategy.

3.3 Trust and Resilience: An Uneasy Alliance

The trust deficits manifest themselves at personal, social, institutional and increasingly virtual levels, where social restraint on online platforms has all but disappeared. Identity politics morph with the disclosure and manipulation of our innermost thoughts (Smee 2018). As the whistle-blower at Cambridge Analytica, the company that exposed the use of data harvested from Facebook accounts, has noted, 'we are trading our democracy for ad optimisation' (Ellison 2018, p. 4). The problem is encapsulated by the fact that

> when you have data and algorithms being weaponised and being used against a population to undermine their perception of what's real . . . our products are, in my view, weapons of mass destruction. And that has been facilitated, in many cases knowingly, by companies like Facebook, that don't see their role as protecting the integrity of our democracy and society, and simply look at it as a platform to make money. (Ellison 2018, p. 4)

When technology becomes a key intermediary in brokering the foundations of what constitutes truth and evidence is distorted in negative feedback loops, the third G30 concern, the bombastic certainty of the 140-character tweet comes into negative play. It offers apparent certainty in an uncertain world. The problem is that it may be wrong. By the time we realise it, the damage has already gone viral, with intended and unintended consequences. The distinction is important and revealing.

The threats to well-being can include natural disasters, such as tsunamis, hurricanes, earthquakes and volcano eruptions. The physical and socio-economic impact can be magnified by disruptive weather patterns associated with climate change and global warming. 'Disaster risk management and adaptation to climate change focus on reducing exposure and vulnerability and increasing resilience to the potential adverse impacts of climate extremes,' for example, is now a key agenda for the Intergovernmental Panel on Climate Change (IPCC 2012). Are the unanswered questions what constitutes effective advance notice? When does the risk appear on the political as well as the meteorological radar? Each constitutes a fact, the management of which constitutes a truth to be tested. Each, however, offers very different versions, with varying consequences for accountability and source of blameworthiness.

Failure to prepare can lead to more extensive flooding or drought. Poor planning decisions across agricultural, semi-rural and urban settings can exacerbate social and physical vulnerabilities. On both spatial and temporal grounds, these mitigation strategies can range from choice of crop production, water conservation policy to economic trade-offs that facilitate provision to build major infrastructure projects on flood plains to accommodate population and economic growth. This is not a problem confined to the developing world as demonstrated by extensive damage in Queensland and its capital, Brisbane, in

2010–2011, which led to insurance payouts totalling $2.38 billion US dollars (Queensland's Floods Commission of Inquiry Final Report 2014, Rec. 4.1a). Houston, Texas, faced similar problems in 2016, as did New Orleans in advance (and more particularly in the aftermath) of Hurricane Katrina in 2007.

Disasters with even larger impacts also occur through direct human action within political and socio-economic spheres. Disruption in one specific domain can prompt contagion. The metastasis occurs most notably, virulently and regularly in finance (Kindleberger 2005; O'Brien 2009; Reinhart and Ragoff 2009; Iwata et al. 2017). Policy calibration to address the problem at the national or international level can, however, produce unintended consequences. Indeed, the innovation may in and of itself provide a transmission belt for further malign metastasis. What works in one jurisdiction may not in another. It may make the patient even more vulnerable. This is particularly the case if those diagnosing and administrating the treatment are not in themselves in agreement on cause and effect or fail to account for the specific social, cultural and political realities at play within the jurisdictions subject to oversight, often without meaningful input or engagement with the populace at large.

The vagueness of the communiqué emanating from the Argentine capital (G20 2018, paras. 24–25) contrasted sharply with direct repudiation of a rise in fuel taxes to encompass a generic sense of grievance in which the richest areas of Paris were targeted for vengeful destruction. The riots that followed were the worst seen in a Western European capital since the imposition of a local council tax in London in 1989. Survey evidence suggests that despite the government blaming the violence on far-left and far-right infiltration, the majority of the French population supports the amorphous movement's aims (*Financial Times* 2018a).

Hope, which is critical to resilience, as a consequence can be undermined, as is the capacity of individuals or nation-states to construct a compelling narrative on how to rebuild in the face of deepening chaos and resentment (Eggerman and Panter-Brick 2010, p. 71) – so too can faith in the promises that short-term collective sacrifice can lay the foundation for sustainable growth if not delivered on or shared evenly. In such circumstances, a perception can take hold that embeds belief in an unidentified yet tangible intellectual elite acceptance of the socialisation of losses and privatisation of profit.

The lexicological shift from risk to resilience is, therefore, not without political calculation (Walker, Holling and Kinzig 2004). On one level, it suggests that lessons have been learnt and that multilateral policy intervention now accounts for the interests of the most vulnerable. Resilience in this sense, like calculation itself, can exude negative connotations (Holling 1973, p. 17). It can privilege the illusion of fundamental change while leaving undisturbed the fundamental causes (Brand and Jax 2007; Welch 2014). Adaptive capacity may mask significant and ongoing imbalances (Fabricus, Foulke, Cundill and Schultz 2007). This is particularly the case in finance (Collier 2018; Deneen 2018; Kuttner 2018). Rebuilding the very system that allowed this to happen turns tragedy into farce.

3.4 The Corruption of Finance

Two unresolved sets of questions remain from the 2007–2008 GFC and its aftermath. The first set focuses on the following: whether and which interests were privileged and at what cost? What sanctions were available to force compliance? Were they used and to what extent have they have served as an effective deterrence? This raises the question of a system's capacity or willingness to adapt when ideational assumptions are undermined, if not altogether falsified.

This fragmentation (as we have seen in relation to the FSB's communiqué to the G20) was evidenced in the fractured relationships within and among the International Monetary Fund, European Commission and European Central Bank troika in the management of the Greek and Irish economies. Following extensive bailouts, invasive conditionality clauses were inserted (and variably implemented). The imposition generated an existential crisis for both debtor nations and creditor institutions. For the nations, it contributed to loss of self-esteem, resentment and the rise of anti-establishment politics (from the left in Greece and through a gaggle of independents in the case of Ireland). For the creditor institutions (European Commission 2017; Fraccaroli 2018; Mody 2018), it exposed the lack of a coherent and cohesive vision in devising a sustained process to harness resources to advance the sustained well-being of the most vulnerable within those countries under economic and social tutelage (Gilmore 2016; Yaroufakis 2017).

These inter- and intra-institutional difficulties also played out in those countries driving the policy agenda. This is particularly the case in the United States. It is a paradox, for example, that the innovation used by the Federal Reserve to provide emergency liquidity at the height of the GFC has now been legislatively prohibited (Posner 2018). The unorthodox action to rescue the global system from its own recklessness and related prior failure of effective supervision became part of the contested narrative over the cause, the source of blame-worthiness and trajectory of reform. Placing such restraints gained immediate political traction in the United States. It may not necessarily help in the event of a future crisis. It may intensify it (Geithner 2017). Seen from this perspective, the question of agency and how to link it to duty and obligation moves centre stage for market regulators and participants alike (O'Brien 2013).

A policy of non-intervention in capital markets, for example, is in itself an interventionist strategy. This is a political choice and one that is normative in nature.

Unless the system as a whole adapts to the changed circumstances – in this case limited room for regulatory intervention – it may in fact make the system less resilient and even more vulnerable, while pandering to closely held but erroneous thinking that the restraints enhances defensive capacity.

The second set of questions, from a socio-economic and political context, is more important. Crime is crime. Breaking the law is breaking the law, even if justice, or its approximation, is dispensed in the civil rather than the criminal court. Even then, infractions do not include admissions of guilt but are subject primarily, through choice, to acceptance of extrajudicial contracts, known as 'deferred prosecution agreements' (O'Brien 2013; Garratt 2014; Rakoff 2019). The criminology literature has long debunked the critical drivers as being individual disposition, greed and non-commitment to commonly held values alone (Ross 1907; Sutherland 1940, 1949; for literature review, see Heath 2008). The same is true in the psychology experiments held in a Stanford University lab and real-world investigations of malpractice in the Abu Ghraib prison (Zimbardo 2007). What then is it that makes the corporate environment, particularly in banking and financial services, so susceptible to criminogenic tendencies?

The manipulation of the London Inter-Bank Offered Rate (LIBOR) and its global facsimiles is a case in point (Talley and Strimling 2013). Within individual institutions, such as Barclays, there was systemic acceptance of misconduct through a policy of neglect (Salz 2013). This facilitated an erosion of personal responsibility across operational risk frameworks, wholesale and retail. The conduct within trading-room subcultures was, however, systemic (O'Brien 2017). It occurred irrespective of ownership structure. Royal Bank of Scotland (RBS) was a repeat offender in both the financial benchmark and Foreign Exchange trading scandals (O'Brien 2017). This occurred notwithstanding the fact that the bank had been effectively nationalised. Neither duty nor obligation to the nation curtailed the misconduct, which was effectively neutralised (or ignored) by senior management (Heath 2008; Admati 2017).

The defence offered by the former head of investment banking at RBS to the British Parliamentary Commission on Banking Standards reflects this normative myopia. He suggested one needed to differentiate between moral bankers and amoral traders, a view endorsed by his colleagues (O'Brien 2014b). The culture thus deified is within a specific community as imagined and as manufactured as the nation-state itself. The downward spiral it facilitates can and does move from self-righteousness and self-pity to delusion. It is remarkable that one of the defences offered by the UBS trader Tom Hayes to his manipulation of the Yen-Libor financial benchmark was that while it may be deemed dishonest by the general public, it would have been seen as acceptable practice within the trading network he worked within. Not surprisingly, the bifurcated test was rejected by the initial court and at appeal. Take, for example, the judgement of Justice Cooke in sentencing Tom Hayes of UBS, the only trader to serve prison time for a scandal of global proportions:

> The essence of your defence was that the type of activity in which you were involved was commonplace in the market at the time and was established practice, not perceived as wrong by those involved. The fact that others were doing the same as you is no excuse, nor is the fact that your immediate managers saw the benefit of what you were doing and condoned it and embraced it, if not encouraged it. ... The conduct involved here must be marked out as dishonest and wrong and a message sent to the world of banking accordingly. The reputation of LIBOR is important to the City as a financial centre and of the banking industry in this country. Probity and honesty are essential, as is trust which is based upon it. The LIBOR activities, in which you played a leading part, put all that in jeopardy. (*R* v *Hayes* 2015, para. 9.12)

Understanding these causes and integrating that knowledge in the search for solutions to strengthen institutions are essential. This need has been demonstrated in empirical research undertaken for the OECD (Caldera Sanchez, de Serres, Gori, Hermansen and Rohn 2016). It suggests that fragility is correlated to the relative strength of financial services in any given economy vis-à-vis countervailing institutional power. It also highlights the importance of articulating and being held to account to a defined purpose. Without it, resilience becomes dangerous. It embeds the very cultures, governance and remuneration that can prove so calamitous to wider society. The imposition of metaphors without accounting for human agency is, therefore, not only misguided. It is counterproductive. And it can lead to ruinous outcomes.

3.5 Turning Architectural Aspiration into Engineering Reality

Ten years after the GFC and the creation of the G20, we have established a global architecture that is fundamentally flawed from an engineering perspective, unlike the famous Sydney Opera House, which encapsulated a vision beyond the existing engineering prowess and required the design of new methods of construction. The iconic physical structure dominates Sydney's harbour despite the begrudgery of short-term politicians focused on playing an anti-elitist populism card and restricting both budget and purpose (Pitt 2018).

By contrast, the G20's approach is one of a tired routine played out to declining and dispirited audiences, in large part because the impresarios of modern politics have a vision as limited as the New South Wales procrastinators or their counterparts in Dublin, determined to rip out the heart of the Georgian city in the name of progress. These disputes are not confined to the past, as shown by the redevelopment of central Sydney, as the state government tears apart local communities to create what the city mayor has slammed as 'ghettoes of the future' (Harris 2018). In this case, those cut out of the redevelopment process are at the lower end of the socio-economic spectrum. Their situation is close to hopeless, shut out from the illusory dreams of a housing folly that replicates and deepens Australia's obsession with real estate. This dynamic corresponds to the findings of ethnographic research across the United States (Hochschild 2016; Vance 2016; Goldstein 2017; Ashwood 2018) and in the United Kingdom (Lanchester 2016; Goodhart 2017; Collier 2018). Along with deprivation and despair, these deep dives reveal a profound sense of loss, distrust in how distant bureaucracies rigidly apply rules and sullen petulance towards what many see as the imposition of liberal values inconsistent with their own. Notwithstanding the alleged absence of class division in the United States, what Pierre Bourdieu (2010) terms the corrosive politics of 'distinction' is played out in a theatre informed by mutual loathing.

The theoretical connection between resilience and sustainability, while often made, is neither necessarily empirically true nor a good thing in itself from a normative perspective (Derissen, Quass and Baumgartner 2011). The utility, efficiency and effectiveness of any given human-driven system cannot be separated from its purpose. The legitimacy and authority of that purpose require the informed consent of those who are subject to it by rules, principles or social norms. What holds the structure together is the dynamic interaction among the three. Reliance on one or two to the detriment of the third risks destabilisation. Each alone has distinct disadvantages. Rules can be transacted around. Principles only work if actually believed in. Social norms are not necessarily socially optimific, especially in cases where the system itself allows by its silence or wilful ignorance breaches of the law to go unpunished (Kahan dnd Braman 2006). The social contract is based on principles of fairness and justice but to remain vibrant requires vigilance to protect the general will against forms of manipulation and self-interest, which, left unattended, can lead inexorably and inevitably to a dystopia of alienation, oppression and unfreedom (Deenan 2018).

These principles require the integration of technical and normative dimensions, which must be adjudicated by the courts (Royal Commission, Final Report 2019). The lack of judicial precedent in the governance and oversight of the financial sector is a causal factor in creating the normative vacuum that led to the GFC. This has been ruthlessly exploited. It has caused untold reputational damage to the national and international orders. It has also rendered corporations and the markets in which they operate exceptionally vulnerable to the politics of opportunism (Levitin 2014; Carney 2015; Lagarde 2017). While it may not be possible to legislate for ethics, it is possible to legislate for breach of promises entered into by virtue of voluntary commitment to legally binding promises (Larson 1977; Kingsford Smith, Clarke and Rodgers 2017). It is essential when designing systems of oversight that these promises are specified. While culture may not be capable of being identified, the results of that culture can be measured (O'Brien and Gilligan 2013; O'Brien et al. 2015).

Metrics must be developed to test relative effectiveness of those promises, with the results (and regulatory responses to them) made public. While considerable work has been undertaken to date to build capital buffers, the work on culture, governance and remuneration, critical drivers of behaviour, has only just begun at an international level and here solely with toolkits for participants and regulators with regards to the wholesale market (e.g. Dudley 2014; Financial Stability Board 2014, 2018b). Even more specifically, sound principles on what constitutes remuneration have long been in existence (Financial Stability Board 2009a, 2009b, 2017, 2018a).

As yet, there has been little attempt to integrate those principles into effective prudential or operational risk management models. One model of resilience could measure persistence or capacity to withstand external shocks without upsetting fundamental power balances. As a consequence, it may elevate the symbolic over the substantive. Another could assess the (expected or unexpected) impact on variegated constituencies. Further parsing could explore the relative power of each to exert or thwart truly transformative outcomes, with or without impacting pre-existing functional utility.

3.6. Disposition, Situation and System

It is essential to note not just how individual parts work but also how the mechanism is held together. In horological terms, we need to understand the dynamics of oscillation within the complex mechanical tourbillon system designed to withstand sudden or incremental build-up of shocks. It is the system itself not just the component parts that is the problem we need to address. Plausible deniability and constructive ambiguity will not suffice. The overt comparison of a social system to an ecosystem, allied to evaluation within the context of adaptation, is part of an appealing but false allure. It has ensured the 'conceptual model has been particularly successful in propagating itself across disciplinary knowledge domains' (Welch, 2014, p. 15).

The essential ambiguity of the very term 'resilience' is both a source of attraction and distraction (Strunz 2012). To understand which is in play at any given time in any given situation requires the analyst to be sensitive not only to the physical landscape but also to how that space is simultaneously patrolled and used as a hunting ground (Welch 2014, p. 21; see also Porter and Ronit 2006). It does not, however, invalidate the concept (Thoren 2014). Abstraction is in and of itself not antithetical to effective public policy. As the history of philosophy shows, the reverse is the case.

If embedded with sufficient cultural sensitivity and nuance, normative frameworks can be established. These can speak to and attract support irrespective of geographic locale or level of educational attainment. With appropriate empirical testing, the bridging of norms and reality as mediated by political and corporate realities can occur. If not, disaster beckons. In an important contribution, Thoren (2014, p. 307) argues that all too often we focus on the 'capacity of a social-ecological system to absorb recurrent disturbances . . . so as to retain essential structures, processes and feedbacks.' All too often, it is this disappointment that serves as the primary driver of the populist turn (Sztomka 1999; Snyder 2018). In this, as in other contexts, resilience may reflect not only failure but also paradoxically success, or more accurately self-interested rather than other-regarding conceptions of what the success means. It also demonstrates that testing must evaluate failure, how to absorb the reasons, communicate meaningful solutions, recover, regroup, adapt and most importantly learn.

In each case, resilience encompasses the physical infrastructure; how information flows within and between institutions at national and international levels; and how, when and how it is disclosed to discrete communities (Linkov, Poinsatte-Jone, Trump, Hynes and Love 2018). None of this is to suggest that it is possible or even desirable to have a zero-tolerance towards risk. What is essential, however, is that the parameters of that risk are mapped and disclosed to secure genuine informed consent. Such an approach builds consensus, reduces fear and controls for anxiety. Politics, like nature, abhors a vacuum. Just because one suggests the need to build resilience into a social system does not necessarily mean it will happen or happen in the way one intends. Indeed, resilience may in fact be what one wants to eradicate. Moreover, weaknesses can breed resentment, not least when accompanied by inability to explain how and why a particular crisis emerged and what can or should be done about it, including the capacity of the judicial system to hold those responsible to account. In the absence of respect, legitimacy and authority, all institutions face the threat of the demagogue.

4 Corporations, Markets and Morals

The GFC was – and remains – the latest, and most catastrophic in recent times, in a series of boom-bust-regulate-deregulate-boom-bust cycles. As the impact moved from the financial into the real economy, the socio-economic costs came into clear view. So too did a failure of policy imagination to prevent, manage or strategically respond (Levitin 2014; O'Brien 2014a). The extent of government intervention required to stabilise financial markets had, for a time-limited period, fundamentally transformed power dynamics. This time, it might, indeed, be different (Reinhart and Rogoff 2009). There was early official recognition of the crisis as a moral one (Brown 2008, 2009; Rudd 2009). This enthusiasm waned, however, the result of panic, fear, necessity, fatigue and the age-old threat of regulatory capture (Landis 1960; Stigler 1971; Posner 1974; Levine and Forrence 1990; O'Brien 2007; Carpenter and Moss 2014). An emphasis on technical solutions to normative problems proved, once again, mistaken.

4.1 Can Technical Measures Address Normative Causes?

Across the banking sector, risk management systems had concentrated on financial not conduct or reputational matters (or wider obligations) when considering operational risk (e.g. PCBS 2013). In determining economic well-being, the main indicator for executives, shareholders and regulators alike was short-term stock performance. Why did it happen? The pithiest analysis comes from the Royal Commission into Misconduct in the Banking, Superannuation and Financial Services Industry (*Final Report* 2019, pp. 1–3). First, 'incentive, bonus and commission schemes throughout the financial services industry have measured sales and profit, but not compliance with the law and proper standards' (*Final Report* 2019, p. 2). Second, 'entities set the terms on which they would deal, consumers often had little detailed knowledge or understanding of the transaction and consumers had next to no power to negotiate the terms' (*Final Report* 2019, p. 2). Third, 'experience shows that conflicts between duty and interest can seldom be managed; self-interest will almost always trump duty' (*Final Report* 2019, p. 3). Fourth, 'too often, financial services entities that broke the law were not properly held to account. . . . Misconduct, especially misconduct that yields profit, is not deterred by requiring those who are found to have done wrong to do no more than pay compensation. And wrongdoing is not denounced by issuing a media release' (*Final Report* 2019, p. 3).

The nexus between misconduct and institutional culture identified is, the report argues, the sole responsibility of the individual corporations involved.

'Entities and individuals acted in the way they did because they could,' Commissioner Hayne posited (*Final Report*, 2019, p. 2). This is the most crucial sentence in the entire report. 'Because They Could' has a powerful emotional resonance. It equates to 'Make America Great', the trope that informs the Trump presidency or 'Take Back Control', the nebulous message that generated a coalition of the discontented to power through Brexit, the 2016 referendum in which the United Kingdom voted to leave the European Union. In the Australian context, 'Because They Could' distils the failure of effective govern-ance across the finance sector and beyond. It encapsulates public perception of a sector whose governance and oversight are informed by neglect and incom-petence, ignorance and opportunity. By no means an Australian-only phenom-enon, the key point of differentiation lies not in Hayne's diagnosis.

As noted earlier, the utility rests in the provision of a route map towards a potentially transformative deliberative exchange. Crucially, this must occur within the finance sector and between it and society, albeit mediated through the political process and determined by the court. It is here that the ambition to reconnect law and morality moves centre stage. For Hayne, this is to be secured within a normative framework operationalised at the corporate level. Specific disputes or departures from this are to be the reserve of future judicial determination.

As such, it reverses the temptation of values being lost in the search for value (MacIntyre 1982, 1984, 1999; Ulrich 2008). While problematic but containable within national domains, the rise of globalism transformed further the dynamics and power imbalances (Stiglitz 2008), nowhere more so than in the world of finance, which gradually mutated from a service into a predatory industry and a classic case study in institutional corruption (Miller 2017, pp. 229–50; Lessig 2018, pp. 29–62). Repairing balance sheets without addressing the underlying condition provided the illusion of change (BIS 2015; Davies and Zhivitskaya 2018). Subsequent misconduct in wholesale market trading of derivative pro-ducts based on compromised financial benchmarks revealed just how resistant these cultures had become. From London to New York, Singapore to Sydney, Paris to Tokyo, traders exploited the naiveté of regulators and politicians alike, creating the populist surge and resentment at a self-serving elite (Judis 2016).

The appropriate first order question, therefore, for corporations and the markets in which they operate is not *how* we regulate or through which mechanism, but *for what specific purpose*? This has the capacity to deliver more effective corporate governance and financial regulation. It is essential, therefore, to evaluate how rules and principles are interpreted within specific communities of practice. There is a dynamic interplay between the culpability of individual actors and the structural and cultural factors that inform their working lives (i.e. how they are compensated; what measures are used to ascertain success or failure; or in more philosophical terms attract merit or approbation, within and beyond their work unit, workplace, competitor corporations as well as broader social circles). The scourge of short-term thinking is as problematic for politics as for markets. Both are socially constructed. Neither has paid sufficient attention to how to embed restraint.

4.2 Rules, Principles and Social Norms

In all cases, three interlinked global phenomena were at play during the GFC: flawed governance mechanisms, including remuneration incentives skewed in favour of short-term profit-taking; flawed models of financing, including, in particular, the dominant originate-and-distribute model of securitisation; and regulatory structures predicated on single institutional failure, not systemic risk. These factors combined to create an architectural blueprint in which innovation trumped security. Financial engineering, in turn, created fiendishly complex products that lacked ethical integrity. Asymmetric information flow, variable capacity – or willingness – to use internal management systems, market mechanisms or regulatory enforcement tools all led to a profound misunderstanding of national and international risks (O'Brien 2009). Deepening market integration ensured that risk, while diversified geographically, remained undiluted. As Nobel Laureate Joseph Stiglitz (2008) stated in excoriating testimony to the US Congress, 'securitisation was based on the premise that a "fool was born every minute." Globalisation meant that there was a global landscape on which they could search for those fools – and they found them everywhere.' In short, financial markets became, as the London-based economist Roger Bootle (2009, p. 239) noted, 'too big, too greedy, too easily drawn to the fabrication of illusory wealth, and too focussed on the distribution of the proceeds, rather than on the financing of wealth creation'. The critical issue is how to deal with a model of capitalism based on technical compliance with narrowly defined legislation and a working assumption that unless a particular action is explicitly proscribed, it is deemed politically and socially acceptable (Sandel 2012).

4.3 Academic Myopia and Lost Legacies

In the search for responsibility and solutions, it is essential that self-reflection extend to the academy, large sections of which ignored how markets can be (and often are) corrupted by a lack of restraint. In each epoch, the social conventions that embed restraint become more ingrained (and specified in law and practice). As Adam Smith (2006, p. 3) puts it in the opening lines of *The Theory of Moral Sentiments*, 'how selfish soever man may be supposed, there are evidently some principles in his nature which interest him in the fortune of others, and render their happiness necessary for him, though he derives nothing from it except the pleasure of seeing it.' In contemporary parlance, empathy replaces sympathy. The other-regarding impulse, however, remains unaltered (Stephan and Finlay 1999; Ulrich 2008, pp. 2–3; Batson and Ahmad 2009; Dovidio et al. 2010). Thus, as Smith (2006, pp. 4–5) argues, 'in every passion of which the mind of man is susceptible, the emotions of the bystander always correspond to what, by bringing the case home to himself, he imagines should be the sentiments of the sufferer.'

For Smith, the same logic applies to our own actions. Restraint, and thus progress and social beneficence, necessitates a confluence of three distinct factors. First, there needs to be personal acceptance of the idealised version of the existing social order but its economic abstraction as Ulrich (2008, p. 2) recognises makes 'mainstream economics ... more a part of the problem than a sound basis for its solution'. Second, ingrained belief that this view corresponds with not only practice but also the belief of others provides trust, authority, confidence and legitimacy. Third, conscience is activated by fear that exposure of deviance from these realities would lead to deserved reputational loss. Not surprisingly, given his staunch Presbyterianism, Smith added a fourth imperative: consequences that extended beyond the grave. Thus, a dread of death and what awaits in the afterlife is 'the great restraint upon the injustice of mankind, which, while it afflicts and mortifies the individual, guards and protects the society' (Smith 2006, p. 7).

Deserved reputation and fear of the shame associated with its loss, therefore, are for Smith critical drivers of individual behaviour and social cohesion. Each holds the other in check because of common understanding of what constitutes propriety. For Smith, this is not a question of manners or social convention. It lies at the core of a deeply held and necessary moral conviction that is believed or perceived to be believed as the very foundation of societal cohesion (Smith 2006, p. 84). Empathy, therefore, acts as a proxy for virtue. Precisely because we are prone to instant gratification, further external supports are, however, also necessary. General prohibitions on causing harm to others are reinforced by positive societal fealty to general rules of individual morality. As society develops, more particularised norms, conventions and laws accrete within an overarching but integrated judicial system to channel and curtail resentment (Smith 2006, p. 76; Frazer 2010, pp. 102–11). Legitimacy and ongoing support depend on the enforceability of a social contract both at law and in social practice, which lies at the heart of the innovation underpinning the Hayne Report in Australia.

All markets have their limits (Sandel 2012). Advanced economic activity, to be sure, cannot take place without the negative restraints of enforceable law.

Stability and beneficence depend, however, on making those who undermine it susceptible to both moral and legal sanctions. Moreover, the transgressions must be actually punished, a key factor animating the Royal Commission. Insofar as there is an invisible hand, it has a distinct other-regarding purpose (Frazer 2010, p. 105). As Smith (2006, p. 160) argued, 'by acting according to the dictates of our moral faculties we necessarily pursue the most effectual means for promoting the happiness of mankind, and may therefore be said, in some sense, to cooperate with the Deity, and to advance as far as in our power, the plan of Providence.'

This connection is not spelt out in *Wealth of Nations*, which concentrates only on the mechanics of economic systems and the minimum requirements for their functioning. There is no irreconcilable conflict between the two works. Understanding the later empirical Smith presupposes awareness of the essential earlier normative reflections of his work as a moral philosopher. There is an ongoing interaction between the prohibitions necessary for economic activity and the positive rules of morality necessary to ensure socially beneficial outcomes. For Smith (2006, p. 87), justice 'is the main pillar that upholds the whole edifice. If it is removed, the great, the immense fabric of human society . . . must, in a moment crumble into atoms'. But this is not deemed sufficient for a healthy polity. Far from it. This nuanced approach to what makes for a peaceful society is lost in accounts that privilege the minimal criteria for economic functionality as presented in *The Wealth of Nations*.

The inherently social basis for Smith's conception of value is embedded throughout *The Wealth of Nations*. As Etzioni (1988, p. 21) has noted, in reading Smith one must consider 'at least two irreducible sources of valuation of utility: pleasure and morality'. Yet this nuance and the social constraints on behaviour are underplayed in classical economics and their misuse is both intellectual conceit and deceit. Likewise, as Andrew Skinner (Smith 1999, pp. 83–84) has eloquently explained, this misreading 'made it possible to think of economics as quite separate from ethics and history, thus obscuring Smith's true purpose. In referring to these problems Terence Hutchison (1988, p. 370), in a telling passage has commented that Adam Smith was unwittingly led by an invisible hand to promote an end result not part of his intention, that 'of establishing political economy as a separate autonomous discipline'. That it did so, notwithstanding Smith's own suspicion of mercantilism, makes the appropriation even more problematic.

The two most significant attempts to control the corporation occurred not in Smith's beloved Scotland but in the United States more than a century later (Johnson 2010). The passage of the Sherman Act (1890) broke the oil and railroad oligopolies central to opening the West. It ended a Gilded Age informed by a logic of freedom to contract and rising inequalities. A subsequent Supreme Court dissent in 1905 in *Lochner* v *New York* struck down the ideational logic associated with freedom of contract but not its practice or its subsequent periodic partial rehabilitation. The Wall Street Crash of 1907, although ruinous, was one that could be contained, but only just. A little more than twenty years later, the rapacious nature of American finance and how it facilitated unconstrained consumption led to the Great Crash and subsequent Great Depression. It delivered to the White House Franklin Delano Roosevelt, the most consequential president of the twentieth century.

Oligopolies were again broken up, this time through the Public Utilities Holding Companies Reform Act (1935). The Securities Act (1933) and the Securities and Exchange Act (1934) put in place the disclosure paradigm that remains the bedrock of financial regulation. Investment and commercial banks were split apart, and the federal government experimented with collective forms of enterprise, such as the Tennessee Valley Authority. The New Deal also signalled the emergence of the regulatory state (Landis 1938), with an uneasy acceptance of regulatory authority conceded with the eventual passage of the Administrative Procedure Act (1946).

As the memories of war receded, a new managerialism dynamic emerged. This reopened the debate on corporate control and purpose. Once seen as decisively won during the New Deal, it turned out to be an uneasy ceasefire. Edward Mason (1958, p. 5) noted the modern corporation had a profound impact on the 'carefully reasoned' laissez-faire defence that 'the economic behaviour promoted and constrained by the institutions of a free market system is, in the main, in the public interest.' For Mason (1958, p. 6), as with Smith before him, this rested on foundations that 'depended largely on the general acceptance of a reasoned justification of the system on moral as well as on political and economic grounds'. The emergence of major corporations, immune from meaningful controls, along with 'apologetics' within the management literature 'appears devastatingly to undermine the intellectual presuppositions of this system' without offering 'an equally satisfying ideology for twentieth century consumption' (Mason, 1958, p. 9). As such, 'the entrepreneur of classical economics has given way to something quite different, and along with him disappears a substantial element in the traditional capitalist apologetic' (Mason 1958, p. 10).

Exactly fifty years later, as the world reeled from the effects of the global financial crisis, Peter Ulrich (2008, p. 3) posed a similar concern that remains essential:

> Normativity always lies behind the economic logic of the market – consequently we have to lay it bare within economic thinking and to reflect upon it in the light of ethical reason. It is important to understand precisely the

thought processes followed in accordance with the inherent logic of the market and to find the hiding places of its normative moments. This allows us to examine the practical (political) programme adopted in its name and to uncover its implicit circumvention of the ethical questions involved in economic action.

It is to this task we now turn.

4.4 The Failure of Law and Economics and the End-of-History Thesis

Despite Mason's misgivings, the economic conception of the corporation as nothing more than a 'nexus of contracts' originally formulated by Coase (1937), but reinvigorated by Friedman (1970; Jensen and Meckling 1976; Fama and Jensen, 1983) spurred on by bombastic protestations from the Chicago School (Friedman, 1970), gradually came to dominate legal scholarship (e.g. Easterbrook and Fischel, 1989, see, also, however, Bratton 1989a, 1989b). Frank Easterbrook and Daniel Fischel, for example, famously argued that wider social issues are and should remain outside market purview or consideration. They cited Adam Smith in defence of the proposition that 'the extended conflict among selfish people produces prices that allocate resources to their most valuable uses' (Easterbrook and Fischel 1989, p. 1422). On this reading, the role of corporate law was solely 'to establish rights among participants in the venture' (Easterbrook and Fischel 1989, p. 1428). For the law and economics scholars, the reduction of complexity to a parsimonious phrasing itself held key normative advantages:

> It removes from the field of interesting questions one that has plagued many writers: what is the goal of the corporation? Is it profit (and for whom)? Social welfare more broadly defined? Is there anything wrong with corporate charity? Should corporations try to maximise profit over the long run or the short run? Our response to such questions is: 'Who Cares?' (Easterbrook and Fischel 1989, p. 1446)

To be effective, however, the model requires suspension of disbelief. It posits adherence to the corporation as a private actor working within efficient markets in which wealth creation provides unalloyed social benefit (Bratton 1989a, 1989b; Orts, 1993). This emaciated, indeed emasculated, approach reached its apogee with the publication of Henry Hansmann and Rainier Kraakman's (2001) landmark essay, 'The End of History for Corporate Law'. The normative claim of 'the end-of-history' thesis was always exceptionally vulnerable to contestation (Greenfield 2006; Stout 2007). The foundational assumption of maximising individual utility cuts against stakeholder and stewardship conceptions of corporate purpose (as well as the philosophical core of Smith's conception of constitutive morality).

The critical challenge is to resolve the existential conflict 'between a public law, regulatory conception of corporate law on the one hand, and a private law, internal perspective on the other' and between 'a body of law concerned solely with the techniques of shareholder wealth maximization [and] a body of law that embraces and seeks to promote a richer array of social and political values' (Millon 1990, p. 202; see also Stone 1981, p. 1442; Collins, 1999, pp. 53–59). President Obama nicely put this conflict at the height of the GFC:

> There's always been a tension between those who place their faith in the invisible hand of the marketplace and those who place more trust in the guiding hand of the government – and that tension isn't a bad thing. It gives rise to healthy debates and creates a dynamism that makes it possible for us to adapt and grow. For we know that markets are not an unalloyed force for either good or for ill. In many ways, our financial system reflects us. In the aggregate of countless independent decisions, we see the potential for creativity – and the potential for abuse. We see the capacity for innovations that make our economy stronger – and for innovations that exploit our economy's weaknesses. We are called upon to put in place those reforms that allow our best qualities to flourish – while keeping those worst traits in check. We're called upon to recognize that the free market is the most powerful generative force for our prosperity – but it is not a free license to ignore the consequences of our actions. (Obama 2009)

4.5 Embedding Integrity through Design

Policymakers and practitioners across the world have acknowledged the pressing need for the development of a regulatory and corporate architecture based on principles of integrity. The zeitgeist has moved decisively from governing to governance, from governance to accountability and from accountability to integrity. If integrity is to have meaning beyond rhetoric, it is essential to parse its multifaceted dimensions as both cause (i.e. its absence) and putative cure for endemic market failure. What does integrity mean in practice? How does one rank potentially incommensurable interpretations of whether behaviour is appropriate? Can one say, for example, that acting within the confines of the law evidences integrity? This surely cannot be a satisfactory answer given the ethical void experienced in both fascist and totalitarian societies, each governed by legal (if morally repugnant) frameworks. Acting within the law does not necessarily equate to ethical behaviour, nor does it provide a foundation for trust in the trustworthiness of others, the essential claim of the legal positivists (Hart 1961). Functional integrity does not necessarily equate to moral integrity. A third, more fruitful approach suggests that propositions of law are true if they figure in or follow from principles of justice, fairness and procedural due process that provide the best constructive interpretation of agreed legal practice.

The scale of ethical failure witnessed in the GFC demonstrates the inherent limitations of black-letter law as a sufficient bulwark. By its very nature, the common law builds on precedent; however, if cases are not brought to judicial determination, there is little to go on. It is equally unsatisfactory to root integrity lexicographically in the application of consistent behaviour. Consistently engaging in deceptive and misleading practice may demonstrate 'wholeness' or 'completeness', but it cannot be a constituent of warranted integrity. Integrity requires of us not only legal obligation (i.e. compliance with the law, consistent and coherent actions). It also requires adherence to broader principles that contribute to (and do not erode) social welfare (i.e. treating people, suppliers and stakeholders with fairness and respect). Seen in this context, enhancing integrity through higher standards of business ethics is a question of organisational design. The aim, in short, is to give substance to what constitutes – or should constitute – appropriate principles of aspiration.

4.6 Navigating the Business Ethics Literature

Business ethics research tends to calcify around one of four main theoretical approaches. These are deontological, consequential or utilitarian, virtue-ethics and contextual based. The deontological approach derives from Immanuel Kant's (1993, p. 30) categorical imperative, namely 'act only according to that maxim whereby you can at the same time will that it should become a universal law.' Reliance on short-term profiteering would, if universalised (and condoned by regulatory and political authorities), destroy the credibility of the market. As such, it is self-defeating. In deontological terms, the GFC evidenced systemic unethical tendencies. Moreover, deceptive or misleading conduct unchallenged at law debased moral capacities.

Viewing the GFC from the less demanding utilitarian perspective, the consequential impact made both the activity itself and the underpinning regulatory framework equally problematic. Here it is essential to differentiate between the nature of a financial product and the uses to which it was put. There is nothing unethical about securitisation per se. However, from an ethical perspective, it is a deficient defence for senior managers to claim ignorance of either how these products were structured or how unstable the expansion of alchemistic engineering had made individual banks (or the system as a whole). It is now recognised, for example, that the originate-distribute-relocate model of financial engineering significantly emaciated corporate responsibility precisely because it distanced institutional actors from the consequences of their actions. Likewise, given the huge social and economic cost, it is insufficient for policy-makers merely to profess shock at the irresponsibility of banks, insurance companies and ratings agencies. The failure to calculate the risks and design or recalibrate restraining mechanisms at the corporate, regulatory and political levels grossly exacerbated the externalities eventually borne by the wider society.

The third major approach, virtue-ethics, is more demanding. It is also more fruitful in terms of refashioning corporate and regulatory action. While the policy response to scandal has traditionally been to emphasise personal character, much less attention has been placed on how corporate, professional, regulatory and political cultures inform, enhance or restrain particular character traits. The focus of virtue-based analysis is on how rules and principles are interpreted in specific corporate, professional or regulatory practice. This is a question of individual and collective character, or integrity. There is prescience to the argument by MacIntyre (1984, p. 196) that the 'elevation of the values of the market to a central social place' risks creating the circumstances in which 'the concept of the virtues might suffer at first attrition and then perhaps something near total effacement.' This builds on earlier insight which suggested that 'effectiveness in organizations is often both the product and the producer of an intense focus on a narrow range of specialized tasks which has as its counterpart blindness to other aspects of one's activity' (MacIntyre 1982, p. 358). Compartmentalisation occurs when 'a distinct sphere of social activity comes to have its own role structure governed by its own specific norms in relative independence of other such spheres' (MacIntyre 1982, p. 358). Within each sphere, those norms dictate which kinds of consideration are to be treated as relevant to decision-making and which are to be excluded (MacIntyre 1999, p. 322). The combination of compartmentalisation and focus on external narrow goods, such as profit maximisation, corrodes the capacity for development of internal goods, which should be developed irrespective of the financial consequences. It is a mistake for institutional economists such as Williamson (2000, p. 597) to assume that social norms, once accreted, remain static.

The search for answers and the putative solutions necessitate that we pay much more attention to the normative dimension of the regulation of capital markets. This, in turn, suggests that regulatory effectiveness cannot be vouch-safed merely by reforming the institutional structure. We need to articulate precisely what is meant by business integrity and accountability within specific contexts. We require a synthesis between an appreciation of context, the need for virtuous behaviour and the importance of deontological rules and consequential principles of best practice within an overarching framework that is not subverted by compartmentalised responsibilities. As such, it is an issue capable of resolution through intelligent application (O'Neill 2002, 2016). Effectiveness needs to be mutually reinforcing and capable of dynamically addressing the calculative, social and normative reasons for behaving in a more (or less) ethically responsible manner (Ulrich 2008).

Despite a rhetorical commitment to enhance integrity, many of the chosen policy options remain firmly within the existing technical realm, relying on traditional regulatory tools such as enhanced disclosure, literacy programs and attempts to distinguish between sophisticated and unsophisticated investors (O'Brien 2014a). Each has proved inadequate in the search for greater or, more accurately, effective accountability. Drawing from O'Neill (2002), it is the opposite of intelligent design.

Given the fact that markets can be and often are inefficient, what is required is an acceptance that effectiveness necessitates dynamic and responsive regulatory guidelines, using the entire suite of enforcement mechanisms, ranging from command and control, through enforced self-regulation, to industry-designed and policed codes of conduct that emphasise social norms. Identifying and repositioning the precise intersection between law and ethics require the design and implementation of an integrated set of nuanced strategies that centre on strengthening internalised belief structures. To be effective, the strategies must align the interests of institutional actors to an overarching regulatory 'mission' or 'purpose', one that recognises the limits of markets and indeed where certain products should be banned irrespective of demand (Sandel 2012; De Grauewe 2017, pp. xiii–xiv).

By building on a foundation of common *stated* values, an agonistic understanding of what constitutes the problematic core is generated, from which deviation from stated value lowers reputational standing and carries punitive sanctions. This framework can only be sustained through an interlocking dissemination network composed of and reinforcing formal and informal nodes. The resulting synthesis has three key practical and normative advantages. First, it reduces real and artificial incommensurability problems between participants in the regulatory conversation (irrespective of whether or not they have been accorded formal surveillance authority). Second, it reduces the retreat to legal formalism, de-escalates confrontation and contributes to behavioural modification across the regulatory matrix. Third, by clarifying accountability responsibilities, it offers greater certainty for corporations and the market in which they are nested, thus facilitating investment flows. It provides a more meaningful baseline from which to measure and evaluate subsequent regulatory and corporate performance. If, as Gordon Brown (2008, 2009) has suggested, there is a need for a new 'social contract' between the banking sector and society at the global level, it is necessary to be much more explicit about the purpose of financial regulation.

5 Contracting Integrity: Legal and Social Licences

The global financial crisis and subsequent scandals, particularly in wholesale markets, have forced renewed critical reflection on the relationship between finance and society. Notwithstanding its periodic and desultory repetitiveness, the core question remains: how to make finance serve society more effectively by better integrating state and market (Polanyi 1944). To shift the parameters decisively, the governor of the Bank of England, Mark Carney (2014), has advocated embedding a normative framework of 'inclusive capitalism'. He argues that higher standards of personal and institutional accountability, linked to a demonstrable rebalancing of liberty and equality of opportunity, will arrest a defined and definable ethical drift within the industry. Moreover, this addresses the heretofore abdication of responsibility by market participants for guaranteeing broader social objectives in exchange for legislative permission to conduct business.

5.1 Constructing Social Obligation

Critical to Carney's thinking is that structural reforms, underpinned by further legislative and regulatory instruments, while essential, are in themselves insufficient. The elusive answer is not more but better regulation, which can and should be internalised at all levels and all stages. Cultural change is, therefore, likely to be more significant in re-engineering the foundations of finance than technical measures alone. Empirical measurement remains essential. It cannot be calculated on financial measures without reference to how short-term profits may compromise external reputation and internal commitment to stated values.

This reformation of values necessitates an articulation and negotiated agreement of a common social purpose. To be transformative, it requires traction at individual, corporate and wider sector levels, determined and overseen by appropriate surveillance mechanisms that are operationally independent but systemically applicable. This ensures that proclamations of probity are, in fact, warranted. Given public scepticism, it is essential that 'soft' regulatory initiatives in the conduct/culture space cannot be dismissed as mere public relations–driven window dressing. Reliance on the banks and politically vulnerable, easily dismissed public officials being the arbiters of obligation is inadequate. It is counterproductive and dangerous given the global populist turn, which tars administrative expertise with the same brush as corporate lobbyists and those they work for (Levine and Forrence 1990; Carpenter and Moss 2014).

The risk of privileging the symbolic applies at each of the building-block levels of belief: basic trust (i.e. inherent), critical trust (i.e. warranted on the basis of a sound operating system that is conceptually and empirically robust) and good-will (i.e. based on belief in the trustworthiness of those to whom basic and critical trust is extended). Each is essential (van't Klooster and Meyer 2015, pp. 9–10). Individuals need to be held to account for their actions/inactions against this trifecta. Critically, so too must institutions. 'Achieving this will require new common standards, cast in clear language, better training and higher qualifications, and ways to ensure that when employees are fired, their history is known to those who consider hiring them,' as Mark Carney (2015) observed in a landmark 2015 speech. Without such an integrated approach, the broader social contract is in danger of breaking down, as 'value becomes abstract and relative. And the pull of the crowd can overwhelm the integrity of the individual' (Carney 2015).

This view of professionalism accords with the idea that values are not inherent in a particular occupational group but how power over that group is exercised (Johnson 1972, p. 45; Larson 1977, p. xvii). It is in keeping with stakeholder theories of sustainable development (Freeman 1984, 2010). It also reflects waning policy support for the shareholder-centric alternative, which has collapsed under the weight of its own contradictions; 2019 is not 2008 and the mores, conventions and pieties of the past view of financial capitalism are no longer fit for purpose (Donaldson and Dunfee 1999). The societal cost of failure has become too large to ignore. If the approach is to become viable as a risk-management tool, institutional strategy and confidence-building measure to ensure commitment to probity and integrity, it is essential to identify and secure support for the most appropriate legal and regulatory structure. This must take cognisance of the fact that the most effective structure may not at any given stage garner public support or trust.

Given the fact that the public is calling for increased accountability, it is no longer credible, for example, for those being called to account to set, monitor and report on the accountability standards alone (although this is an essential component). At every stage, therefore, one has to adjudicate an appropriate balance within and between self-regulation, co-regulation and mandated approaches. All count for nothing, however, if the agenda represents the triumph of the politics of symbolism. Sturdier supports are essential. The commitments must be intrinsic (i.e. the underpinning values that anchor the social compact must be internalised and believed by specific communities of practice and consistent with public expectation).

The aim is not to impose a culture template. Rather, it is to ensure that espoused culture is lived, that confidence in stated commitments is warranted. If viewed as an encumbrance, the entire artifice crumbles. It is important to note that reform initiatives that help industry inculcate restraint and safeguard the integrity of markets are a long-standing if not necessarily remembered goal of regulatory intervention. Seen in this context, the 'inclusive capitalism' initiative spearheaded by the Bank of England reinserts a lost normative dimension (Carney 2014, 2015), similar to a repositioning at the OECD towards a new emphasis on guiding markets in the direction of inclusive economic growth leading to better lives. Similar changes in rhetoric are also apparent in discrete policy circles in the United States (Dudley 2014). In Australia, the idea of a 'social licence to operate' (SLTO) has received overt (if sporadic) political traction. It sputtered to a halt during a particularly bad-tempered debate centred on a proposed rewording of the ASX Principles of Corporate Governance (2018, principle 3). Although the final version jettisoned the reference to a social licence in favour of measures to identify, enhance and protect corporate reputation through acting with integrity, it remains a latent if not potent force for change. It is in the United Kingdom, however, that one sees the most extensive and exhaustive articulation of how the restoration of lost legitimacy, credibility and trust serves defined public goods:

> Markets need to retain the consent of society – a social licence – to be allowed to operate, innovate and grow. Repeated episodes of misconduct (such as the Libor and FX scandals) have called that social licence into question. To restore it, we need to rebuild fair and effective markets. Not markets that collapse when there is a shock from abroad. Not markets where transactions occur in chat rooms. Not markets where no one appears accountable for anything. Real markets are professional and open, not informal and clubby. Their participants compete on merit rather than collude online. Real markets are resilient, fair and effective. They maintain their social licence. (Carney 2015)

The program, sketched by the thoughtfully structured British Parliamentary Review of Banking Standards report *Changing Banking for Good* (2013), was set out two years later in the *Fair and Effective Markets Review* (2015). The FEMR clarifies at high-level principle the duties and responsibilities of senior management. It overhauls the structure of some markets. There is an equal emphasis on the need for soft governance mechanisms for embedding cultural change. As before, there is also an explicit invitation to industry to further develop and implement defined higher standards, rather than mandating the imposition of a one-size-fits-all model. Designed and implemented with conviction, the framework offers a route map to deliver enlightened self-interest, linked to long-term sustainability that is owned by industry itself and which industry has an obligation to uphold. It mandates comprehensive coverage and the development of credible sanctions, with industry-wide applicability. As such, it offers a strengthened form of 'meta-regulation' through which political and regulatory authorities can and indeed *should*, as representative of societal interests, play a framing role. If not gamed, this approach remains the 'smartest' and most effective form of regulation.

What is also clear is that the inherently secretive nature of prudential regulation means that operationalising the framework can only take place through a market-conduct prism and then only through application of broader frameworks that emphasise responsibility and commitment. The financial regulation agenda remains conceived in self-referential terms. It needs to tie more directly to the United Nations and OECD frameworks on corporate responsibility and their relationship with human rights protection, a reduction of tax arbitrage and direct and indirect corruption. This is a critical failing. These global protocols provide a solid basis for the evaluation of a growing regulatory and public policy imperative. They serve five interlinked purposes. They (1) reduce creative compliance, (2) provide for warranted commitment to high ethical standards and legitimate institutional purposes, (3) enhance the effectiveness of deterrence, (4) improve accountability and (5) reduce institutional and systemic risk. In so doing, they address the very problems and solutions identified by industry but not necessarily followed through as a consequence of heretofore suboptimal design. It is in this context that the question of culture remains material (and missing).

5.2 Mediating the Culture Wars

Culture mediates competing values at multiple levels and multiple domains. It is a process of knowledge creation, socialisation and legitimation. It is an instrument of power and control as much as freedom of expression. Culture can be empowering. It can also be malign. It is far from being a neutral concept or an aspirational goal. Culture needs to be understood before it can be evaluated. We need to know much more about how its values are created and framed. This process is by no means communicated at the level of universal application. Differential schooling can be embedded within stratified higher education systems and solidify further within ever more constrained, and constraining, communities of practice (Bourdieu 1996, 2010). This stratification is by no means unique to Bourdieu's investigation of metropolitan Paris. The mores of metropolitan elites generate and allow for a heuristic framing that transcends specific cultures.

Cultural standards, once adopted and followed, reflect the clarity and the myopia within both individual institutions and sectors as a whole (be they in finance or university humanities faculties). These 'culture wars' can be diagnosed and benchmarked in business (Rossouw and van Vurren 2003). What should unite all approaches is civility and respectful curiosity (Brooks 2016), a dynamic and reflexive differentiation and resolution among three core competing dynamics within modern capitalist societies. It is rarely present because of these dynamics.

These are, first, the relative privileging of asceticism over ostentatious acquisition; second, acceptance of bourgeois society values or elevation of the moral relativism and accompanying cynicism of modernism; and, third, and crucially, the integration or separation of law and morality (Bell 1996, pp. 293–85). In each case, high culture has become more fractured, if not unanchored, from underpinning societies and values. Postmodernism, as the French sociologist Jean Baudrillard (1976) has complained, has facilitated a blurring. This makes the real hard to distinguish from the media construction of reality. As Baudrillard (1994, p. 15) puts it, 'capitalism in fact, was never linked by a contract to the society that it dominates. It is a sorcery of social relations, it is a challenge to society and it must be responded to as such.' For Baudrillard, technological advances accentuate a core paradox. The expansion of access to information is accompanied by a reduction in experienced meaning, associated with a lack of depth, coherence, meaning, authenticity and originality. It is the latter that has become, by far, the more pronounced and consequentially problematic. The consensus on separation of law and morality and concomitant elevation of the transactional over the relational has become more homogenous among competing elites. Criticising those who bemoan the economic, social and political devastation caused by globalisation as misanthropes may play well in parts of Islington, Saint-Germain, Sydney's Eastern Suburbs or Manhattan's Upper West Side.

Far removed from the epicentre of the largest-scale migration shifts since the Second World War, with gentrification a symbol of faux connection to artisan pasts, these global denizens live and work within a media and social bubble. In reality, these suburbs are composed of not one but two or perhaps 'multi-elites', one defined by education, the second by both accrued wealth and income. One is cultured, the second mercantile (Piketty 2018, p. 4). Elites have proved incapable of understanding never mind addressing the existential angst of those rendered invisible except as servicing agents. The significance of Piketty's observation is that across each of the countries surveyed, the future is indeterminate, with multiple cleavages possible. As he puts it, '"multiple-elite" stabilization; complete realignment of the party system along a "globalists" (high-education, high-income) vs "nativists" (low-education, low-income) cleavage; return to class-based redistributive conflict, either from an internationalist or nativist perspective' (Piketty 2018, p. 3).

In sharp contrast to the self-referential political and economic critiques that preordained neoliberalism as both cause and cure, Carney was among the first to recognise the need to reach out in a genuine partnership to the disaffected. It is

> vital that we – public authorities and private market participants – work together to reverse the tide of ethical drift. This cannot be a one-off exercise; we need continuous engagement so that market infrastructure keeps pace with market innovation. In so doing it is possible to build the real markets the UK deserves. Markets that merit social licence and reinforce social capital. (Carney 2015)

For Carney, the reflection must start at the level of the corporation itself and extend through the markets in which it trades.

The creation, testing and evaluation of how actual conduct and (prior) generally accepted business practice impact internal risk management and external compliance with regulatory objectives are essential. They can provide evidence of the disconnection between stated and lived obligation. It may shift the boundaries of acceptable private practice by fusing hard and soft law with voluntary initiatives, thus addressing public unease. It is equally the case that in the absence of effective policing of these agreed indicators, engagement may give credence to a system that flatters to deceive. Risk of failure, however, should not invalidate investigation. It does, however, require an understanding of how the socialisation of ideationally driven narratives occurs. This, in turn, necessitates examining how and why corporate law and financial regulation discourse turned on procedural rather than substantive matters. It was, as we shall see, not accidental.

The core governance problems surrounding corporate purpose were crystallised in the seminal volume *The Corporation in Modern Society* (Mason 1960). For Mason (1960, p. 1), 'to suggest a drastic change in the scope or character of corporate activity is to suggest a drastic alteration in the structure of society. . . . All of this is to suggest not that the corporation cannot be touched but that to touch the corporation deeply is to touch much else beside.' Notwithstanding this observation, the debate on corporate law and governance, their form and purpose, can be reducible to two main schools of thought. The first canvasses communitarian approaches to governance. It posits that the corporation has responsibilities to protect social welfare as well as rights (Mitchell 1995; Stout

2007). These inform the normative assumptions that underpin the Mason volume. The second privileges a freedom to contract model. Crucially, this leaves it to the corporation to decide the optimal balance between narrow self-interest and societal obligation (Coffee 1989). The distinction does not obviate the necessity to integrate technocratic and normative dimensions. Non-intervention in the internal governance of the firm in itself is an interventionist position, a critical point of contention in regulatory theory and practice (Landis 1938; Ogus 1994; O'Brien 2014). It also privileges a normative stance (and the economic interests of those to benefit the most), as Hayek (1943, p. 29) understood only too well. It also represented the triumph of an ideational framework, which obfuscated the political calculation required to construct what was presented as the economically rational (Hansmann and Kraakman 2001; see also, however, Ireland 2000; Greenfield 2006). It was in essence a political hijack (Ulrich 2008).

5.3 Constructing a Social Licence

So just what is the social licence? How can it be negotiated, implemented, monitored and evaluated? Who should be responsible? What lessons can be learnt from its application in other sectors, particularly the extractive industries, where it originated as a short-term risk management issue? How and when does a corporation transcend bottom-line or indeed triple-bottom-line reporting – economic, social and environmental – and on what basis?

Given the business judgement rule, which leaves it to the corporation itself to decide the form and content of internal controls within broadly defined legal frameworks, how does one account for variation in internal dynamics, management styles and authority structures? Moreover, good governance cannot rely solely on negative sanctions alone, formal or informal. Embedding, reinforcing, institutionalizing the SLTO require mechanisms that are themselves accountable, proportionate, reasonable and fair. It is essential, therefore, to generate a coherent and cohesive holistic framework that evaluates cultural mores across six core dimensions: purpose, philosophy, parameters, principles, policy and process.

At the outset, it is essential to note the limits of the SLTO concept. Extending obligation where there are no regulatory requirements to reach or maintain consent can be problematic. It may be difficult to identify relevant stakeholders or rank competing views. Without a legislative or listing obligation underpinning, the social licence is a voluntary initiative designed to demonstrate good faith and, crucially, build trust – hence, its practicality and allure. It is a description of the pressure facing a business in the event that it loses public support and legitimacy, prompting the business risk of formal rejection of a specific project or a requirement for more invasive oversight. It is also a model for effective community engagement and how to attain and/or retain legitimacy, credibility and trust (Boutilier and Thomson 2011; Thompson and Boutilier 2011). As such, it can be an effective risk management tool (King and Ruggie 2005; Porter and Kramer 2006).

The model has its origins in World Bank research on sustainable development. It refers to the capacity of a business to conduct its affairs outside of a government-sanctioned or legally enforced process (Wilburn and Wilburn 2011). The World Bank has noted that community legitimacy is based on securing the 'free, prior and informed consent of local stakeholders', a requirement of a prior United Nations initiative to support the rights of indigenous people (United Nations 2004). The concept has been developed most extensively in the extractive industries arena where managing social risk is very much an ongoing business risk issue (Henisz, Dorobantu and Nartey 2014). It is indicative that in a 2012 Ernst & Young (2012) report, the ability to retain the SLTO was ranked the fourth most significant risk facing mining. In 2014, it was rated third and in 2015 fifth.

The problems for the mining companies are exacerbated by 'digital disruption' (intensifying and compressing the news cycle) and the fact that the term has entered the public consciousness through unmoderated private social media communication channels, increasingly incapable of tracking or rebutting (King and Soule 2007). By 2018, it had dropped to seventh place, although a new criterion of 'regulatory risk' caused as a consequence of 'governments and regulators in developing countries have intensified their focus on implementing new laws aimed at greater local participation which has brought uncertainty and risk to the sector' and moved the concept of conduct risk to number five (Ernst & Young 2018, p. 6).

Data shows that the costs of failing to pay attention to SLTO risk are substantial (David and Franks 2014, p. 8) Along with direct and indirect immediate financial costs are reputational ones (McCormack and Stears 2014). These can, of course, have more significant long-term implications than headline-grabbing fines and compensation payments. In addition, when crisis management dominates the agenda and time of senior executives, attention is necessarily distracted from strategic priorities. Moreover, increased litigation risk (and willingness to bluff) increases legal defence fees and settlement costs. As Ernst & Young has noted,

> while there are no set guidelines on what steps a company needs to take in order
> to obtain or maintain its SLTO, it is increasingly clear that very early engagement

in employing a collaborative, trust-based model that includes effective engagement with stakeholders will achieve a greater level of credibility, a stronger sense
of legitimacy and acceptance, and a healthier legacy than anything a formal
license can offer. (Cited in David and Franks 2014, p. 18)

5.4 Lessons from Human Rights Discourse

Significant progress in achieving that goal was contained in a report to the
United Nations in 2010 setting out how the state could shape industry preferences through policy coherence and recommending that

> incentives to communicate adequate information could include provisions
> to give weight to such self-reporting in the event of any judicial or
> administrative proceeding. A requirement to communicate can be parti-
> cularly appropriate where the nature of business operations or operating
> contexts pose a significant risk to human rights. Policies or laws in this
> area can usefully clarify what and how businesses should communicate,
> helping to ensure both the accessibility and accuracy of communications.
> (Ruggie 2010, p. 9)

Professor Ruggie goes on to argue for vertical and horizontal policy
coherence:

> Vertical policy coherence entails States having the necessary policies, laws
> and processes to implement their international human rights law obligations.
> Horizontal policy coherence means supporting and equipping departments
> and agencies, at both the national and sub- national levels, that shape business
> practices – including those responsible for corporate law and securities
> regulation, investment, export credit and insurance, trade and labour – to be
> informed of and act in a manner compatible with the Governments' human
> rights obligations. (Ruggie 2010, p. 12)

Now consultancies have mushroomed, setting out handbooks on how to manage
individual crises. What has not been attempted before is applying the model
across an entire sector. What is envisaged by calling for a social licence for
finance, therefore, is nothing less than a fundamental repositioning of a critical
industry.

It forces the sector to engage in meaningful and sustained dialogue with
its stakeholders about the risks, benefits and impacts of its collective
decisions (as one might expect, for example, from an airline or car manufacturer, anxious to reassure customers about their personal safety in the
aftermath of accident or scandal). Unless resourced and genuinely independent, there is a real danger that the process will fail, calcifying cynicism
and escalating the downward spiral into distrust. To avoid this happening,

the social licence model requires detailed evaluation of who the stake-holders are and what responsibilities the sector has to them. It requires transparent and accountable decision-making processes, constant monitoring and adaptation in light of changing circumstances. In short, it needs a fully funded holistic research agenda.

This is essential at this point because sustained unprofessional, self-interested and harmful behaviour within the finance sector has caused such damage to wider society. There is an urgent need to change the operational culture to restore to an equitable position the balance between the privileged participation and potential for rewards as licenced financial services actors that individuals and organisations receive and the civic duties and obligations that could, and indeed should, accompany that privileged status. Although significant progress has been made in establishing a market standards board for the fixed income currency and commodities (FICC) sector, its remit appears at this stage too limited to provide this core function. Moreover, the other professional bodies, or those aspiring to such status, remain too disparate to provide warranted confidence in the placing of trust in the sector as a whole. Unlike a single accident in a single country, or a single social dispute over a single mining operation, the problems facing the finance sector are systemic and global. Rebuilding trust requires systematic evaluation.

The moral failings of the market have been a defining feature of myriad official inquiries into the GFC. The British Parliamentary Commission on Banking Standards (2013) has carried out the most detailed evaluation of ethical deficits and whether these could be addressed by importing systematically into finance professional norms and mores. From the beginning, the commission identified a major problem. The professionalisation project presupposed that there existed within the capital markets a distinct kind of activity that could be characterised as having the attributes of professional life (e.g. specific tertiary educational requirements that act as a barrier to entry; ongoing professional development; meaningful codes of conduct that are effectively monitored and enforced; effective and demonstrable commitment to the development and enhancement of professional standards; and, crucially, mechanisms to suspend or withdraw a professional licence to operate in the event of misconduct). Notwithstanding the stated commitment of the British Banking Association of the need for a professional body with requisite regulatory power, the final report of the Parliamentary Commission demonstrated acute wariness. Banking, it concluded, 'is a long way from being an industry where professional duties to customers, and to the integrity of the profession as a whole, trump an individual's own behavioural incentives' (PCBS 2013, vol. 2, para. 596). The Hayne Royal Commission came to a similar conclusion.

5.5 The Ties That Bind: My Word Is My Bond

Taken together, these steps suggest that evaluation cannot be conducted by the banks or entities which are beholden to them. In any case, banks (and regulators) tend to be reluctant to be the 'first mover' even when the need for change has become apparent. It is not so much that they do not wish to reform themselves, more that they would rather someone else did it first (especially if the needed reform involves major cultural change at the potential expense of short-term profits). What is needed is a circuit breaker, a truly independent entity that acts as an evaluative agency and an honest broker, consistently staging a series of events to ensure constant levels of oversight and capacity to introduce and promote and advocate small steps that work. Such an approach may lack the allure of the Guildhall. It may have the attraction, however, of lowering the social and business costs and risks of finance. It is a shrewd investment.

Existing codes of conduct at the corporate, industry or professional level proved incapable of addressing hubris, myopia and the decoupling of ethical considerations from core business. The failure to articulate and integrate purpose, values and principles within a functioning ethical framework created or exacerbated socially harmful corporate cultures. These cultures elevated technical compliance over substance. Ethical obligation was stated but not delivered. Deterrence was defective and ineffective. There was little or no accountability. No credible mechanisms to identify institutional or systemic risk were put in place.

While the structural reform agenda has been significantly advanced and the social licence to operate model shows much promise, there is a very real danger that the old lies of finance – that this time is different, that markets always clear and that markets are moral – will retain their power over the incredulous. Carney (2015) has warned that 'to resist their siren calls, policymakers and market participants must bind ourselves to the mast. That means building institutional structures that make it harder to act on the lies.' There is little point, however, going into a storm on a boat not fit for its purpose.

6 Reconnecting Law and Morality through Principle

On the tenth anniversary of the GFC, we were treated to a veritable avalanche of coverage. The financial media concentrated more on remembrance than analysis of current risk. What that actually means in practice is very much open to question. *The Economist* (2018a, 2018b) noted despite the scarifying nature of the GFC, structural change was minimal in the eye-of-the storm of 2007–2008, and in response to misfeasance and malfeasance in wholesale markets post bailouts: 'If it is time to believe that the crisis that began in 2008 has really ended, it is past time to wonder how the new conditions which have come about in its wake could contribute to the one that comes next' (The *Economist* 2018a, pp. 20–22). Enhanced supervision was piecemeal. Regulatory responses lacked granular inspection of particular risk. It is a point reluctantly conceded by *The Economist* which reminded its international readership of an ideal lost in translation: 'The true spirit of liberalism is not self-preserving but radical and disruptive' (*The Economist* 2018b, p. 17). Radicalism and disruption are, however, like middle-class recruits to contemporary killing fields, missing in action.

Amid the war stories and the recreation of investor panic (and handsome recovery), there was no serious consideration of those most vulnerable to the near collapse of the global economy. The fate of those who lost their homes and how they had fared over the intervening decade were unexamined. The social cost in terms of lost services, opioid epidemics, psychiatric illnesses and individual and familial stress and breakdown was not calculated. The cultural and political factors that created the consensus which dislodged the political from a uniform managerial consensus on the triumph of markets were unexplored (Westbrook 2010; Braithwaite 2013; O'Brien and Gilligan 2013).

As we have seen throughout this Element, this consensus permeated a flawed building process that misunderstood or misrepresented purpose (Afsharipour 2017, p. 467). It ranged from suboptimal design to construction and certification on what were unstable foundations (Frank 2001, pp. 15–23; O'Brien 2003, 2007, 2009). We must also reflect, however, on the implications of past blind faith in markets, current reflexive anti-establishmentarian impulses and the risks of dismantling the international rules-based order. For this evaluation to begin, we must ascertain the actual panoply of social, political, economic and critically cultural variables at play, then and now, within and between national markets. The relationships among the self, culture and civilisation are inextricably intertwined. It is a complex dynamic exchange that is determined by form, process and custom.

The core of potential conflict rests on the dichotomy between conceiving humankind as a collection of individuals or a social species. If the former, does individual obligation extend beyond voluntary self-restraint, informed by the exercise of individually relative moral freewill? If the latter, what boundaries can or should be placed on individual desire? How could and should these be justified? When do the structures of civilisation become prisons rather than secular cathedrals to celebrate and venerate shared values and meanings? As theorist Terry Eagleton (2016, p. 26) acidly pointed out, 'desire scoops out a hollow in humanity, overshadowing presence with absence and spurring us beyond the given to whatever eludes our grasp. In this sense, it can be seen as the very dynamic of civilised existence.' For Eagleton (2016, p. 26), desire itself has no moral compass and is directionless; as such, 'it signifies a flaw at the very heart of our fulfilment, an errancy of our being, a homelessness of the spirit.'

By any reckoning, this is a bleak view of progress. It does, however, highlight that left to our own devices, we cannot be sure our mechanical pursuit of desire will lead us to socially beneficial outcomes. As Eagleton chides with character-istic acuity, those seeking help from psychotherapy rarely get it from self-awareness but from engagement within a defined theatre operating with observed and observable roles. By its very nature, the ego is not constrained or necessarily aware of its own impulses. Moreover, it is likely to seek self-rationalisation to avoid or deflect responsibility.

From the affluent campus of Stanford University in 1971 (Zimbardo 2007, p. x) to the cave of Lydia in the myth of what constitutes restraint by Glaucon to Socrates in Plato's *The Republic*, behavioural psychologists and ancient philo-sophers point in the same direction of travel. Restlessness can be a curse as well as a blessing. The pursuit of acquisition, the triumph of individualism and the division of law and morality are, as Daniel Bell (1996, pp. 283–85) reminds us, the core contradictions at the heart of capitalism. These contradictions are not new, but the ordering within an overarching framework of analysis is excep-tionally useful. It is necessary, therefore, to ascertain the relative, contingent and deliberated balance of the rights of individuals to and from the society in which they reside. To gain legitimacy and authority, this reasoning must be transpar-ent, capable of review and follow due process. Concomitantly, it is necessary to link the individual to the communitarian underpinnings by exploring the rela-tive role played by fairness (or in a legal sense principles of equity) and efficiency (normally but not exclusively calculated by cost considerations alone or if at all in philosophical terms by principles of utility).

A further, increasingly contested and problematic area of general concern is how to define the parameters of what constitutes the private. This includes determining whether and how informed consent has been obtained and what sanctions could or should be imposed for breaking the written or unwritten principles governing the polity. These core issues lie at the heart of the unful-filled social contract imagined by Jean Jacques Rousseau, who, in lamenting the possibility of securing stability felt himself drawn to the politics of the strong-man, as we are again today. A public philosophy integrating law and morality remains elusive, not least because of the decline in transcendent belief.

In an increasingly secular age, we are witnessing a shift from what Taylor (2007, p. 713) calls 'hierarchical mediated-access societies to horizontal, direct access societies. And second, the modern social imaginary no longer sees the greater trans-local entities – nations, states, churches – as grounded in something other, something higher, than common action in secular time.' Paradoxically, this freedom may lead to disenchantment and nostalgia for the recreation of some kind of order (Taylor 2007, p. 714). The politics of populism are best seen as a form of disappointment, and short-term disappointment has been inbuilt to the mainframe of chrematistic logic since Aristotle (Daly and Cobb 1994, p. 138). None of this is to deny the moral value of identity but to situate it within a larger collective than used (and abused in contemporary politics on both the left and the right).

The challenge is epitomised by the triumph of sloganeering. 'There is no such thing as society,' Margaret Thatcher was once fond of reciting. Nowhere was this preference for parsimonious phrasing over evaluation more emblematic than in the most consequential referendum in the history of the United Kingdom of Great Britain and Northern Ireland. 'Take back control' in the 2016 Brexit campaign for the UK to leave the European Union without a deal or route map for that deal went beyond individual agency. It simultaneously manipulated the politics of nostalgia. It harked back to a world in which civilisation was accompanied by national control over boundaries and destinies (and their export). Pride can, however, come before a fall, especially when emotion trumps calculation in political games. When the winning hand appears to result only in the retention of the status quo, longer-term implications begin to emerge. The Brexit bet displayed this chrematistic recklessness. Erroneously, it was viewed in tactical terms as a single transaction. There was little regard for either the long-term impact or cost-benefit evaluation for the broader community. Issues like climate change make this much more difficult to quantify and evaluate.

Third, the launch into the unknown, fuelled only by hope, was one with little practical use or certainty. For Taylor (2007), it is a malaise that has dominated reasoning since the age of the Enlightenment and the rise of secularism itself:

> Running through all of these attacks is the spectre of meaningless; that as a result of the denial of transcendence, of heroism, of deep feeling, we are left with a view of human life which is empty, cannot inspire commitment, offers nothing really worthwhile, cannot answer the craving for goals we can dedicate ourselves to. Human happiness can only inspire us when we have to fight against the forces which are destroying it; but once realised, it will inspire nothing but ennui, a cosmic yawn. (Taylor 2007, p. 717)

What emerges when this normative backdrop is applied to the finance sector is how very particular and what a narrow form of culture and accompanying particularised logic it encapsulates. In this context, culture and its commodification become part of the problem itself (Eagleton 2016, p. 151). This is socialised and given authority and standing according to self-referential but exclusive rather than inclusive terms (Bourdiou 1996). Paradoxically, it is one that is also evident in the City of London. Still driven by class distinctions, moral bankers are contrasted to immoral traders. Place both in the same toxic environment, however, and it leads, inexorably, to the triumph of moral relativism. This occurs precisely because of the lack of an underpinning purpose or indeed philosophical grounding (Bell 1996, p. 252). It is a fact acknowledged by the independent review into the culture of Barclays, commissioned by the bank after the outbreak of the LIBOR scandal:

> Transforming the culture will require a new sense of purpose beyond the need to perform financially. It will require establishing shared values, supported by a code of conduct, that create a foundation for improving behaviours while accommodating the particular characteristics of the bank's different businesses. It will require a public commitment, with clear milestones and regular reporting on progress. It will require Barclays to listen to stakeholders, serve its customers and clients well, get on with the work to implement its plans and stay out of trouble. The complexity of Barclays' businesses makes this a particular challenge for its leaders. It will take time before it is clear that sustainable change is being achieved. (Salz Review 2013, para. 2.20)

This very process makes 'culture' such a malleable and dangerous emotional term. It can signify belonging. It can also promote a sense of rightfulness and nostalgic camaraderie that can rapidly disintegrate. This downward spiral can and does move from self-righteousness to self-pity to delusion. Capital remains global, while its regulation is national (or in the case of the European Union regional and therefore more cohesive, but prone to accusations of being unaccountable or privileging the interests of its strongest members; see Varoufakis 2017; Mody 2018). Capital remains unanchored from the societies in which transnational corporations operate, and while vulnerable to disruption to supply chains caused by imposition of tariffs, this collateral damage is less worrisome than legislative restrictions on future use of the very innovative techniques that minimised the collateral damage of the banking sector's recklessness.

The opening salvoes of a global trade war now reverberate on skittish emerging markets. Populist resistance across the political spectrum from Argentina to Brazil is fuelled by the search for scapegoats. The risk of capital flight intensifies as the international rules-based order buckles under the weight of intemperate rhetoric emanating from its Washington, DC, core. Protectionism may lead to short-term electoral success. It certainly has for Donald Trump, who, during the 2016 presidential campaign, told a rally that steelmaking jobs losses were a 'politician-made disaster', the 'consequence of a leadership class that worships globalism over Americanism' (Woodward 2018, p.134). It was and remains a good line. It did and does little to secure protection from a dangerous blowback derived from a failure to address the root normative causes of the GFC or indeed its major antecedent, the collapse of the dot.com bubble in the United States or debt defaults in both Asia and Russia, the latter inextricably linked to the 1998 collapse of Long Term Capital Management, a notorious hedge fund whose collapse necessitated a collective bailout from Wall Street itself. The sheer scale of the GFC and how the logic and practice of financial capitalism had become so intertwined made a private sector solution an impossibility.

These variables exist within a dynamic space. Latitude within it is predetermined by the previous balance of economic power and actual capacity to commodify it in political terms. The exercise of that power is not determined by the strength of parliamentary democracies alone. Across political systems, from the democratic to the authoritarian, capacity is framed by multifaceted terms of reference. These set the parameters of debate and their evaluation. They constitute and limit challenge to any given system's hegemony: its legitimacy, its authority and its durability. In adopting this much more granular approach, it is essential to disentangle the wiring of hegemonic systems. We must map and evaluate how perceptions, values, beliefs and social norms become weaponised. These are deployed within what the Italian Marxist theorist Antonio Gramsci (1971) terms ongoing wars of position and wars of manoeuvre. These conflicts have psychological as well as geographic dimensions.

The battlefield, metaphorically speaking, is located across global locales, with malware as effective as military drones in destroying carefully cultivated ecosystems. In the cyberworld, as in conventional military planning, tactics and strategy combine to privilege hybridity. Those seeking to attain, retain or challenge hegemony nurture and prune cultural framing. The results come into bloom at differential times and not necessarily in an integrated or identifiable form. Plausible deniability remains very much in demand in corporate and political manipulation. The innovation can capture the imagination or animate boredom and cynicism. On this somewhat pessimistic reading, as in our instant acceptance of codes of conduct and service in accessing software updates, it is Thrasymachus not Socrates who describes morality more accurately as it actually is: a cover for power, an argument more forcefully articulated two millennia later by Nietzsche and why nationalism is the preserve of the populist.

The true genius of nationalism was to find common ground. The 'master concept' for fertilisation was and remains identity politics. What constitutes identity, however, has become increasingly difficult to ascertain. It is also difficult to agree or prioritise or delegate authority for the solution of urgent commonalities to third-party agents. What becomes lost in translation within multicultural, often segregated (not in legal form but in social substance) de-industrialised society is what holds us together. The problem for the left is that in having cultivated identity politics, it has ceded mastery to the right.

In a secular age, neither religion nor personal commitment to individual virtues holds collective persuasive power or sway. One can hold progressive views on social matters but be conservative in economic ones. The opposite is also the case. In each, however, the combination of economic distress and perceived loss of dignity has become fertile ground for those who argue general collective concerns are discounted, leading to an increase in 'invisibility' as a key driver of dissatisfaction and alienation. At a more benign level, across the world social experience has itself become atomised, even at times of profound common interest, for example, a sporting final, a riot, a terrorist attack or a royal wedding. Not only do we increasingly experience these things alone. We communicate our feelings about them through (voluntarily closed yet ostensibly open) networks. These are designed primarily to facilitate confirmatory bias.

The GFC and how it was experienced is a further case in point (as indeed is the resulting opioid crisis, which now kills more people per annum in the United States than fatalities caused by traffic accidents). Outside the financial arena, the tenth anniversary rarely featured in the mainstream and social media, other than a historical curiosity. In part, this reflects the editorial ordering of self-curation (reinforced by the popularity editorial ranking of mainstream press). In part, it also demonstrates how a retreat to technicalities in search for solutions itself plays a role in hegemonic battles. The GFC and its causes have been forgotten. The strength of a hegemonic power rests in its power to anesthetise. We remain under the care of the surgeon's knife – in politics, so too, if not more so, in regulation. It has become the core arena for civilian political conflict in the modern administrative state (O'Brien 2014a).

Corporate, political, regulatory and judicial cycles spin at different speeds, with differing indicators of success. The problem intensifies with complexity. Parsimony does not necessarily provide the substance of sustainable reform. It can elevate the politics and practice of symbolism. The prime cause of the crisis came not simply from individual greed. Misaligned incentives within and across banking institutions contributed to the inculcation of a collective, if reckless, worldview but are insufficient. Nor is it the case that rashness can be attributed solely to the fact that these institutions were so interconnected that they were deemed 'too big to fail'. Bailouts or guarantees were neither inevitable nor indeed in themselves knowable to work in advance. They almost destroyed Ireland, which has made a steady if slow recovery, now compromised by the reality of Brexit. They contributed to delayed onset of cultural symptoms in Australia. It is fair to say each of these factors combined to light the bonfire of the vanities.

The cross-pollination facilitated a fundamental misjudgement of the parameters and public tolerance of risk at operational and reputational levels and its impact on perceptions of dignity and respect. Everyone was convinced everyone knew what they were doing. None in a position of power stopped to consider consequences on those vulnerable to a dilution of respect and dignity. Profit-making departments trusted their innovators and the capacity of traders to operate within sound ethical boundaries. Institutions trusted the efficacy of compliance systems, most notably the 'three lines of defence' model, which had received implicit imprimatur from the Basel Committee (Davies and Zhivitskaya 2018). Regulators, particularly on the prudential side, confused the model's existence with assurance (from themselves or those who deployed it) that it either worked or had even been tested. Overarching this entire structure was political nonchalance. This went far beyond quiescence. This, indeed, occurred. It included the sanguine acceptance of risk in pursuing deregulatory agendas. There was a profound lack of curiosity about the increasing complexity of finance. Early warning signs were ignored, even when they came from the regulators themselves. Mandates were, at best, incomplete. None showed any real interest in testing the limits of law or application of policy. The processes of accounting left accountability and responsibility if not ignored then certainly not prioritised. In large part, this reflected the consequences of conflating conflicting policy imperatives (which proved too electorally successful and too suicidal to ignore).

Expanding home ownership was a bipartisan goal particularly in the United States and the liberalisation of markets was one sure way of facilitating the voluntary assumption of unsustainable levels of debt. Likewise, there is nothing intrinsically unethical in the alchemy of securitisation (in either real or synthetic forms, although the latter is more akin to casino gambling). If there was a perfect storm, it was one powered by more than hubris. Ignorance played a more decisive role. At times, that ignorance was wilful. As a consequence, it is no wonder that faith in the trustworthiness of our institutions has withered.

Trust in the financial system to put the interests of anyone or anything beyond its own narrow self-interest has been damaged. As a consequence, the values of banking have come under clear scrutiny. The dire warning underpins the interim and final findings of the Royal Commission into Misconduct in the Banking, Superannuation and Financial Services Industry (*Final Report* 2018, 2019). The Deloitte 'Trust in Banking' survey provides statistical analysis aligned with Hayne's investigation. It also provides targeted areas of focus for each aspect of the sector, regulated entities and their regulators to repair the damage caused by misconduct.

The theory and practice of corporate governance and financial regulation are out of alignment. Failure to address this existential point exacerbates the problem. It is the system itself not component parts that is the problem, a finding consistent with the academic literature on applied ethics and criminology. The downward spiral of trust in banks as institutions and their oversight, both legal and regulatory, represents a massive challenge for Australia's financial services industry. This is not a crisis that can be fixed by empty rhetoric or meaningless spin.

The precariat is no longer confined to working-class outer suburbs, or class distinctions. Political affiliation is no longer relevant as a key point of differentiation, until one moves outside of the mainstream. Legitimacy and authority cannot survive without trust. Paradoxically, both the problem and the solution to it reflect the weakness of the state in mediating the interdependent relationship between the finance sector, a reliance on market-based solutions and the complicity of the political establishment.

A reliance on a rules-based approach runs the risk of being transacted around. Countervailing preference on a principles-driven approach lacks traction if unaccompanied by actual belief in those other-regarding restraints. The distrust is not just in the products and services but also in the underlining purpose behind them. This can offer opportunity for both clients and the banks themselves to break free from the straitjacket of apathy if the corporate mindset is one of demonstrable keeping of promises and the products and services can be differentiated on ethical grounds.

All societies are held together by a combination of rules, principles and social norms. Each element requires constant affirmation and reaffirmation to be held in creative and productive tension. Commissioner Hayne in Australia is to be applauded for demanding the active use of the judiciary as honest brokers and definitive guardians of the symbiotic relationship between law and morality. His castigation of the market conduct and prudential regulator in not testing the law enough is couched very carefully. Hayne's essential point is that this systemic political failure could and should be challenged.

This is not to say that active testing of the law will necessarily produce a more restrained market. Nonetheless, it is also clear that the non-calculative social contract that underpins legal, governance and transactional activity has been breached. It requires renegotiation to ensure substantive not technical compliance, warranted not stated commitment to high ethical standards, effective and measurable deterrence and enhanced levels of personal and institutional accountability to ensure risk frameworks reflect living cultures. It is not difficult. It does, however, need to reflect on the dangers of interdependence and the privileging of compartmentalisation between law and morality.

At board level, a standing committee, akin to the Ethics Committee of the Norwegian Government Pension Fund Global, the world's largest sovereign wealth fund that benchmarks financial performance against ethical conduct, is a start. Notwithstanding the scepticism, and the realisation that cultural change is likely to be generational, a start must be made. Extensive polling data, a moral compass and navigational route can lead to delivery of the OECD's new mantra that the purpose of markets and corporations listed within them is to facilitate 'inclusive economic growth leading to better lives'. It is a journey worth pursuing not least because the insertion of normative imperatives such as 'inclusive' and 'better' are far removed from the utility wealth-maximising principles of previous fealty to efficiency alone.

To be effective, it must be one informed by far more extensive and independent analysis and review of proposals than heretofore provided. Cost-benefit analysis must, like remuneration strategies, include non-financial measurements. The dialogue must be curated and sustained within defined holistic parameters and based on verified and verifiable information. This is not to suggest that a solution be imposed. It is to say that to be effective as a true exercise in deliberative governance, purpose and effectiveness must act as proxies for fairness and efficacy. Evaluation cannot be undertaken from within the system itself. The system, with cause, cannot be trusted.

Challenge and opportunity in a post-truth, post-Bannon world do not lie in amplifying the siren call of lost tradition or finding scapegoats within failed bureaucracies. In each case, the administrative state is presented as a self-interested 'deep state'. Lost in this narrative is any tangible way to improve it. Australia is no different. Paradoxically, both the problem and the solution to it reflect the weakness of the state in mediating the interdependent relationship between the finance sector, a reliance on markets and the political establishment. What the Deloitte/Trust project finds is distrust across the board not just in the products and services but also the underlining purpose behind them. The distrust rises the more informed and the more money one earns. There is an equal lack of faith in regulatory authorities to manage an interdependence that is both toxic and proven to be inherently unstable. Trust lost has to be earned back – by actions, not words.

References

Admati, A 2017, 'It takes a village to maintain a dangerous financial system', in L Herzog (ed.), *Just financial markets: finance in a just society*, Oxford University Press, Oxford.

Adorno, T 1950, *The authoritarian personality*, Harper & Brother, New York.

Afsharipour, A 2017, 'Redefining corporate purpose: an international perspective', *Seattle University Law Review*, vol. 40, pp. 465–96.

Agius, M et al. 2010, 'Financial leaders pledge excellence and integrity,' *Financial Times*, 29 September.

Arendt, H 2017, *The origins of totalitarianism*, Penguin Classics, London.

Aristotle, 2004, The *Nicomachean ethics*, Penguin Classics, London.

Arndorfer, I & Minto, A 2015, 'The "four lines of model defence" for financial institutions', Bank for International Settlements, Occasional Paper Number 11, December.

Ashwood, L 2018, *For-profit democracy: why the government is losing the trust of rural America*, Yale University Press, New Haven.

Awrey, D & Kershaw, D 2014, 'Towards a more ethical culture in finance: regulatory and governance strategies' in N. Morris and D. Vines (eds.), *Capital failure: rebuilding trust in financial services*, Oxford University Press, Oxford.

Ayres I & Braithwaite, J 1992, *Responsive regulation*, Clarendon Press, Oxford.

Baars G & Spicer, A 2017, *The corporation: a critical multi-disciplinary handbook*, Cambridge University Press, Cambridge.

Bakan, J 2005, *The corporation: the pathological pursuit of profit and power*, Free Press, New York.

Bank for International Settlements 2015, *Basel committee on banking supervision corporate governance principles for Banks*, BIS, Basel.

Barker, R (ed.) 2011, *Corporate governance, competition, and political parties: explaining corporate governance change in Europe*, Oxford University Press, New York.

Batson C & Ahmad, N 2009, 'Using empathy to improve intergroup attitudes and relations', *Social Issues and Policy Review*, vol. 31, no. 1, pp. 141–77.

Baudrillard, J 1994, *Simulcra and simulation*, University of Michigan Press, Ann Harbor.

1976, *Symbolic exchange and death*, Sage Publications, London.

Bell, D 1996, *The cultural contradictions of capitalism*, Basic Books, New York.

Berle, A 1931, 'Corporate powers as powers in trust', *Harvard Law Review*, vol. 44, no. 7, pp. 1049–74.

Means, G 1932, *The modern corporation and private property*, MacMillan, New York.

Bicchieri, C 2017 *Norms in the wild: how to diagnose, measure and change social norms*, Oxford University Press, New York.

Blair, M & Stout, L 1999, 'A team production theory of corporate law', *Virginia Law Review*, vol. 85, no. 2, pp. 247–328.

Bootle, R 2009, *The trouble with markets: saving capitalism from itself*, Hachette, London.

Botsman, R 2017, *Who can you trust?* Penguin, London.

Bourdieu, P 2010, *Distinction*, Routledge, London.

1996, *The state nobility: elite schools in the field of power*, Polity Press, Cambridge.

1990, *The logic of practice*, Stanford University Press, Stanford, 2000.

Boutilier, R & Thomson, I 2011, 'Modelling and measuring the social licence to operate: fruits of a dialogue between theory and practice,' *Social Licence*. Available at https://socialicense.com/publications/Modelling%20and%20Measuring%20the%20SLO.pdf

Braithwaite J 2013, 'Cultures of redemptive finance', in J O'Brien and G Gilligan (eds.), *Integrity risk and accountability in capital markets: regulating culture*, Hart Publishing, Oxford.

Brand, F & Jax, K 2007, 'Focusing the meaning(s) of resilience: resilience as a descriptive concept and a boundary object', *Ecology and Society*, vol. 12 no. 1, p. 23 (online).

Bratton W 1989a, 'The nexus of contracts corporation: a critical appraisal', *Cornell Law Review*, vol. 74, no. 1, pp. 407–65.

1989b, 'The new economic theory of the firm: critical perspectives from history' *Stanford Law Review* vol. 41, July, pp. 1471–1527.

Wachter M 2008, 'Shareholder primacy's corporatist origins: Adolf Berle and the Modern Corporation', *Journal of Corporation Law*, vol. 34, no. 1, pp. 99–152.

Bremmer, I 2018, *Us vs. them: the failure of globalism*, Penguin, London.

Brooks, D 2016, 'Let's have a better culture war', *New York Times*, 7 June.

Brown, G 2009, 'Speech to G20 finance ministers', St Andrews, Scotland, 7 November.

2008, 'The global economy', Speech delivered at the Reuters Building, London, 13 October.

Brown K & Westaway, E 2011, 'Agency, capacity, and resilience to environmental change: lessons from human development, well-being and disasters', *Annual Review of Environment and Resources*, vol. 36, pp. 321–42.

Caldera Sánchez, F, de Serres, A, Gori, F, Hermansen M & Röhn O 2016, 'Strengthening economic resilience: insights from the post-1970 record of severe recessions and financial crises', *OECD Economic Policy Papers*, 20, OECD Publishing, Paris.

Carney, M 2018, 'Letter to G20 Leaders from FSB chair', Financial Stability Board, 27 November.

2017, 'Worthy of trust: law, ethics and culture in banking', Speech delivered at the Bank of England Conference Centre, London, 21 March.

2015, 'Three truths for finance', Speech delivered at the Harvard Club UK, Southwark Cathedral, London, 21 September.

2014, 'Inclusive capitalism: creating a sense of the systemic', Speech delivered at the Inclusive Capitalism Conference, London, 27 May.

Carpenter, D & Moss, D (eds.), 2014, *Preventing regulatory capture: special interest influence and how to influence it*, Cambridge University Press, New York.

Cassidy, J 2009, *How markets fail: the logic of economic calamities*, Farrar, Strauss and Giroux, New York.

Clarke, T, O'Brien, J & O'Kelley, C (eds.) 2019, *The Oxford handbook on the corporation*, Oxford University Press, Oxford.

Coase, R 1937, 'The nature of the firm', *Economica*, vol. 4, November, pp. 386–405.

Coffee, J 1989, 'The mandatory/enabling balance in corporate law: an essay on the judicial role'. *Columbia Law Review*, vol. 89, pp. 1618–91.

Cohen, N 2017, *The rise of Silicon Valley as a powerhouse and social wrecking ball*, One World, London.

Collier, P 2018, *The future of capitalism: facing the new anxieties*, Penguin, London.

Collins, H 1999, *Regulating contracts*, Oxford University Press, Oxford.

Daly, H & Cobb, J 1994, *For the common good: redirecting the economy towards community, the environment, and a sustainable future*, Beacon Press, Boston.

David, R & Franks, D 2014, 'Costs of company-community conflict in the extractive industry', Corporate Social Responsibility Initiative Research Report 66, Harvard Kennedy School.

Davies, H & Zhivitskaya, M 2018, 'Three lines of defence: a robust organising framework, or just lines in the sand?' *Global Policy*, vol. 9, supplement 1, June, pp. 34–42.

De Grauwe, P 2017, *The limits of the market: the pendulum between government and market*, Oxford University Press, Oxford.

Deenan, P 2018, *Why liberalism failed*, Yale University Press, New Haven.

Denzau, A & North, D 1994, 'Shared mental models: ideologies and institutions', *Kyklos* vol. 47, February, pp.3–31.

Der Spiegel, 2017, 'Trump and Bannon pursue a vision of autocracy', 6 February.

Derissen, S, Quaas, M & Baumgärtner, S 2011, 'The relationship between resilience and sustainability of ecological-economic systems', *Ecological Economics*, vol. 70, no. 6, pp. 1121–28.

Dobel, JP 1999, *Public integrity*, Johns Hopkins University Press, Baltimore.

Dodd, EM 1932, 'For whom are corporate managers trustees', *Harvard Law Review*, vol. 45, no. 7, pp. 1145–63.

Donaldson, T & Dunfee, T 1999, *Ties that bind: a social contracts approach to business ethics*, Harvard Business School Press, Boston.

Dovidio, J, Johnson, J, Gaertner, S, Pearson, A, Saguy, T & Asburn-Nardo, L 2010, 'Empathy and inter- group relations', in M. Mikulincer and P. Shaver (eds.), *Prosocial motives, emotions, and behaviour: the better angels of our nature*, APA Books, Washington, DC.

Downs, A 1957, *An economic theory of democracy*, HarperCollins, New York.

Dudley, W 2016, 'Opening remarks at reforming culture and behaviour in the financial services industry: expanding the dialogue', Speech delivered at the Federal Reserve Bank of New York, New York, 16 October.

2014, 'Enhancing financial stability by improving culture in the financial services industry', Speech delivered at Federal Reserve Bank of New York, New York, 20 October.

Eagleton, T 2016, *Culture*, Yale University Press, New Haven.

Easterbrook, F & Fischel, D 1989, 'The corporate contract', *Columbia Law Review*, vol. 89, pp. 1416–48.

Edelman Trust Barometer 2012–2019, Edelman Publications, New York.

Economist, The 2018a, Briefing, The financial crisis: unresolved', 8 September, pp. 20–22.

2018b Editorial, 'Has finance been fixed', 8 September, p. 11.

Eggerman M & Panter-Brick, C 2010, 'Suffering, hope and entrapment: resilience and cultural values in Afghanistan', *Social Science and Medicine*, vol. 71, no. 1, pp. 71–83.

Elliot, TS 1948, *Notes towards the definition of a culture*, Faber & Faber, London.

Ellison, J 2018, 'Christopher Wylie: what happens next', Life and Arts, Financial Times (Asia Edition). 1–2 December, pp. 4–5.

Ellison, S 2012, 'Intelligent accountability: rethinking the concept of "accountability" in the popular discourse of education policy' *Journal of Thought*, vol. 47, no 2, pp. 19–41.

Ernst & Young 2012, *Business Risks Facing Mining and Metals*, Author, New York.

Ernst & Young 2018, *Business Risks Facing Mining and Metals*, Author, New York.

Etzioni, A. 1988, *The moral dimension: towards a new economics*, The Free Press, New York.

European Commission 2016, 'On state aid, SA.38373 (2014/C) (ex 2014/NN) (ex 2014/CP)', Implemented by Ireland on Apple, Brussels, 30 August.

European Commission Staff Working Document 2017, 'Coping with the international financial crisis at the national level in the European context', *Impact and Financial Sector Policy Responses 2008–2015* SWD (2017) 373 European Commission, Brussels, 21 November.

Fabricius, C, Foulke, C, Cundill G & Schultz, L 2007, 'Powerless spectators, coping actors, and adaptive co-managers: a synthesis of the role of communities in ecosystem management', *Ecology and Society*, vol. 12, no. 1, p. 29 (online).

Fama, E & Jensen, M 1983, 'Separation of ownership and control', *Law and Economics* vol. 26, no. 2, pp. 301–25.

Financial Times 2018a, 'The French pick up their pitchforks against president Macron', Editorial, 3 December, p. 8.

2018b 'US needs multilateralism as much as its partners', Editorial, 29 December, p. 6.

Financial Stability Board 2018a, *Implementation and effects of the G20 financial regulatory reforms fourth annual report*, Basel, 28 November.

2018b, *Supplementary guidance to the FSB principles and standards on sound compensation standards: the use of compensation tools to address misconduct risk*, Basel, 9 March.

2017, *Implementing the FSB principles for sound compensation practices and their implementation standards, fifth progress report*, Basel, 4 July.

2014, *Guidance on supervisory interaction with financial institutions on risk culture: a framework for assessing risk culture*, Basel, 7 April.

2009a, *Principles for sound compensation standards practices*, Basel.

2009b, *Principles for sound compensation implementation standards*, Basel.

Fligstein N & Dauter, L 2007, 'The sociology of markets', *Annual Review of Sociology*, vol. 33, no. 6, pp. 1–24.

Foreign Affairs 2018, Editorial, 'Which world are we living in', July/August, pp. 7–8.

Four Corners 2018, 'Populist revolution', ABC Television (Australia), 3 September.

Fraccaroli, N, Giovannini, A & Jamet, JF 2018, 'The evolution of the ECB's accountability practices during the crisis', *European Central Bank Economic Bulletin*, Issue 5.

Frank, T 2001, *One market under god: extreme capitalism, market populism and the end of economic democracy*, Anchor Books, New York.

Frazer, M 2010, *The enlightenment of sympathy: justice and moral sentiments in the eighteenth century and today*, Oxford University Press, New York.

Freeman, R 2010, *Strategic management: a stakeholder approach*, Cambridge University Press, New York.

    1984, *Stakeholder management: framework and philosophy*, Pitman, Mansfield, MA.

Friedman, M 1970, 'The social responsibility of business is to increase its profits', *New York Times Magazine*, 13 September, pp.32–33, 122–26.

Fukuyama, F 1995, *Trust: the social virtues and the creation of prosperity*, Free Press, New York.

G20 2018, *G20 Leaders' Communiqué: Building Consensus for Fair and Sustainable Development*, Buenos Aires, 1 December.

    Framework Working Group 2017a, 'Note on resilience principles in G20 countries', Baden-Baden, 18 March.

    2017b, *G20 Hamburg action plan*, Hamburg, 8 July.

G20/OECD 2015, *G20/OECD principles of corporate governance*, OECD Publishing, Paris.

G30 2018, *Banking conduct and culture: a permanent mindset change*, G30 Publications, Washington, DC.

Garratt, B 2014, *Too big to jail: how prosecutors compromise with corporations*, Harvard University Press, Cambridge, MA.

Geithner, T 2017, 'Are we safe yet', *Foreign Affairs*, January/February.

Gilmore, E 2016, *Inside the room: the untold story of Ireland's crisis government*, Merrion Press, Dublin.

Goldstein, A. 2017, *Janesville: an American story*, Simon & Schuster, New York.

Goodhart, D 2017, *The road to somewhere: the new tribes shaping British politics*, Penguin, London.

Gorbachev, M 2017, *The new Russia*, Polity Press, Cambridge.

2009, 'Time for a global perestroika,' *Washington Post*, 7 June.

Gramsci, A 1971, *Selections from the Prison Notebooks*, International Publishers, New York.

Greenfield, K 2006, *The failure of corporate law: fundamental flaws and progressive possibilities*, University of Chicago Press, Chicago.

Greenspan, A 2008, Evidence to the House Committee on Oversight and Government Reform, Washington, DC, 23 October.

Halperin, JL 2011, 'Law in books and law in action: the problem of legal change', *Maine Law Review*, vol. 64, no. 1, pp. 46–76.

Hansmann, H & Kraakman, R 2001, 'The end of history for corporate law', *Georgetown Law Review*, vol. 89, pp. 439–68.

Harris, J 2018, 'Ghettoes of the future: Waterloo housing proposal slammed', *Architectureau*, 8 August.

Hart, H 1961, *The concept of law*, Oxford University Press, Oxford.

Hayek, F 1943, *The road to serfdom*, Routledge, London.

Heaney, S 1998, 'Scaffolding', *100 Poems*, Faber & Faber, London.

Heath, J 2008,'Business ethics and moral motivation: a criminological perspective', *Journal of Business Ethics*, vol. 83, no. 4, pp. 595–614.

Henisz, W, Dorobantu S & Nartey L 2014, 'Spinning gold: the financial returns to stakeholder engagement', *Strategic Management Journal*, vol. 35, no. 12, pp. 1727–48.

Hochschild, A 2016, *Strangers in their own land: anger and mourning on the American right*, New Press, New York.

Holling, C 1973, 'Resilience and stability of ecological systems', *Annual Review of Ecology and Systematics*, vol. 4, pp. 1–23.

Hosking, G 2014, *Trust: a history*, Oxford University Press, Oxford.

Inglehart, R 2018, 'The age of insecurity: can democracy save itself?' *Foreign Affairs*, May.

2018, *Cultural evolution*, Cambridge University Press, New York.

IPCC 2012, Managing the risks of extreme events and disasters to advance climate change adaptation, http://ipcc-wg2.gov/SREX/report/

Ireland, P 2000, 'Defending the rentier: corporate theory and the reprivatisation of the public company', in J Parkinson, A Gamble and G Kelly (eds.), *The political economy of the company*, Hart Publishing, Oxford.

Iwata, K, Jean, S, Kastrop, C., Loewald, C & Veron, N 2017, 'Resilience and inclusive growth', *G20 Insights*, 24 May.

Jensen, M & Meckling, W 1976, 'Theory of the firm: managerial behaviour, agency costs and ownership structure', *Journal of Financial Economics*, vol. 3, no. 4, pp. 305–60.

Johnson, N 2017, 'The rise of the precariat', *Economia*, 2 June.

Johnson, P 2010, *Making the market: the Victorian origins of corporate capitalism*, Cambridge University Press, Cambridge.

Johnson, T 1972, *Professions and power*, Macmillan, London.

Judis, J. 2016, *The populist explosion: how the great recession transformed American and European politics*, Columbia Global Reports, New York.

Kahan D & Braman, D 2006, 'Cultural cognition and public policy', *Yale Law and Policy Review*, vol. 24, no. 1, pp. 147–72

Kahneman, D 2012, *Thinking, fast and slow*, Penguin, London.

Kakutani, M 2018, *The death of truth*, HarperCollins, London.

Kant, I 1993, *Grounding for the metaphysics of morals*, Penguin, London.

Kavanagh, J & Rich, M 2018, *Truth decay: an initial exploration of the diminishing role of facts and analysis in American public life*, Rand Corporation, Santa Monica, CA.

Keasey, K, Short, H & Wright, M 2005, 'The development of corporate governance codes in the United Kingdom,' in K Keasey, S Thompson & M Wright (eds), *Corporate governance: accountability, enterprise and international comparisons*, John Wiley & Sons, Chichester.

Kindleberger, C 2005, *Manias, panics and crashes: a history of financial crises*, John Wiley & Sons, Hoboken, NJ.

King, B & Ruggie, J 2005, *Corporate social responsibility as risk management: a model for multinationals*, Corporate Social Responsibility Initiative Research Paper 10, Kennedy School of Government, Harvard University, Cambridge, MA.

Soule, S 2007, 'Social movements as extra-institutional entrepreneurs: the effects of protests on stock price returns', *Administrative Science Quarterly*, vol. 52, no. 3, pp. 413–42.

Kingsford-Smith, D, Clarke T & Rogers, J 2017, 'Banking and the limits of professionalism', *University of New South Wales Law Journal*, vol. 40, no 1, pp. 411–55.

Kuttner, R 2018, *Can democracy survive global capitalism?* Norton, New York.

Lagarde, C 2017, 'A time to repair the roof', Speech delivered at JFK School of Government, Harvard University, 5 October.

Lanchester, J 2016, 'Brexit Blues', *London Review of Books*, 15 July, pp. 3–6.

Landis, J 1960, Report on regulatory agencies to the president elect, Washington DC, 21 December.

1938, *The administrative process*, Yale University Press, New Haven.

Larson, M 1977, *The rise of professionalism: a sociological analysis*, University of California Press, Berkeley.

Lenin, V 1902, *What is to be done? burning questions for our movement*, Foreign Languages Press, Peking, 1973.

Lessig, L 2018, *America compromised: five studies in institutional corruption*, Chicago University Press, Chicago.

Levine, M & Forrence, J 1990, 'Regulatory capture: public interest and the public agenda: towards a synthesis', *Journal of Law, Economics and Organization*, vol. 6, April, pp. 167–98.

Levitin, A 2014, 'The politics of financial regulation and the regulation of financial politics: a review essay', *Harvard Law Review*, vol. 127, no. 1, pp. 46–55.

Linkov, I, Poinsatte-Jone, K, Trump B, Hynes, W & Love P 2018, 'Resilience at the OECD: current state and future directions', Working Paper, SG/NAEC (2018) 5, JT03435565, OECD, Paris, September.

MacIntyre, A 1999, 'Social structures and their threats to moral agency', *Philosophy*, vol. 74, no. 789, pp. 311–29

1984, *After virtue: a study in moral theory*, University of Notre Dame Press, South Bend.

1982, 'Why are the problems of business ethics insoluble,' in B Baumrin & B Friedman (eds.), *Moral responsibility and the professions*, University of Michigan Press, Ann Arbor.

Mansbridge, J 1999, 'Altruistic trust', in M Warren (ed.), *Democracy and trust*, Cambridge University Press, Cambridge.

Mason, E 1958, 'The apologetics of managerialism', *Journal of Business*, vol. 31, no. 1, pp. 1–11.

1960, 'Introduction,' in E Mason (ed.), *The corporation in modern society*, Harvard University Press, Cambridge, MA.

Mathews Burwell, S 2018, 'Generation stress: the mental health crisis on campus', *Foreign Affairs*, November.

McCormack, R & Steers, C 2014, 'Banks: conduct costs, cultural issues and steps towards professionalism', *Law and Financial Markets Review*, vol. 8, no. 2, pp. 133–44.

Mendelson, E 2016, 'In the depths of the digital age,' *New York Review of Books*, 23 June.

Miller, S 2017, *Institutional corruption: a study in applied philosophy*, Cambridge University Press, Cambridge.

Millon, D 1995, 'Communitarianism in corporate law: foundations and law reform strategies', in L Mitchell (ed), *Progressive corporate law*, Westview Press, Boulder, CO.

1990, 'Theories of the corporation', *Duke Law Journal* vol. 39, no. 2, pp. 201–62.

Mitchell, L (ed.) 1995, *Progressive corporate law*, Westview Press, Boulder, CO.

Mody, A 2018, *EuroTragedy: a drama in nine acts*, Oxford University Press, Oxford.

Moorhead, R & Hinchley, V 2015, 'Professional minimalism: the ethical consciousness of commercial lawyers', *Journal of Law and Society*, vol. 42, no. 3, pp. 387–412.

Mouffe, C 2018, *For a left populism*, Verso, London.

Muller, JW 2016, *What is populism?* University of Pennsylvania Press, Philadelphia.

Nichols, T. 2017, *The death of expertise: the campaign against established knowledge and why it matters*, Oxford University Press, New York.

Norman, J 2018 *Adam Smith: how he thought and why it matters*, Penguin, London.

Obama, B 2009, 'Remarks by the president on 21st century financial regulatory reform', Speech delivered at Press Conference, White House, Washington DC, 17 June.

O'Brien, J 2017, 'The FX global code: transcending symbolism', *Law and Financial Markets Review*, vol. 11, nos. 2–3, pp. 83–95.

2016 'Shooting fish in a barrel: investor protection in the aftermath of the GFC', *Law and Financial Markets Review*, vol., 10, no.3, pp. 117–22.

2014a, *The triumph, tragedy and lost legacy of James M Landis: a life on fire*, Hart Publishing, Oxford.

2014b, 'Professional obligation, ethical awareness and capital market regulation' in N Morris and D Vines (eds.), *Capital failure*, Oxford University Press, Oxford.

2013, 'The façade of enforcement: Goldman Sachs, negotiated prosecution and the politics of blame', in S Will, D Brotherton and S Handelman (eds.), *How they got away with it: lessons from the financial meltdown*, Columbia University Press, New York.

2009, *Engineering a financial bloodbath*, Imperial College Press, London.

2007, *Redesigning financial regulation: the politics of enforcement*, John Wiley & Sons, Chichester.

2003, *Wall Street on Trial*, John Wiley & Sons, Chichester.

Gilligan, G 2013, 'Regulating culture: problems and perspectives,' in J. O'Brien and G. Gilligan (eds.), *Integrity, risk and accountability in capital markets: regulating culture*, Hart Publishing, Oxford.

Gilligan G, Roberts, A & McCormick, R 2015, 'Professional standards and the social licence to operate: a panacea for finance or an exercise in symbolism', *Law and Financial Markets Review* vol. 9, no 4, pp. 283–92.

OECD 2018, *Flexibility and proportionality in corporate governance*, OECD Publishing, Paris.

Ogus, O 1994, *Regulation: legal form and economic theory*, Hart Publishing, Oxford.

O'Neill, O 2016, 'What is banking for', Speech delivered at the Federal Reserve Bank of New York, New York, 16 October.

    2002, *A question of trust: the BBC Reith lectures 2002*, Cambridge University Press, Cambridge.

Orts, E 1993, 'The complexity and legitimacy of corporate law', *Washington & Lee Law Review*, vol. 50, pp. 1565–623.

Parfit, D 2011, *On what matters*, Oxford University Press, Oxford.

Parliamentary Commission on Banking Standards (PCBS) 2013, *Changing banking for good*, HM Treasury London.

Picciotto, S 2011, *Regulating global corporate capitalism*, Cambridge University Press, Cambridge.

Piketty, T 2017, *Capital in the twenty-first century*, Harvard University Press, Cambridge, MA.

    2018 'Brahmin left vs merchant right: rising inequality and the changing structure of political conflict: evidence from France, Britain and the US, 1948–1970', World Inequality Database, Working Paper 2018/7, March.

Pitt, H 2018. *The house: The dramatic story of the Sydney Opera House and the people who made it*, Allen & Unwin, Sydney.

Polanyi, K 1944, *The great transformation: the political and economic origins of our time*, Farrar & Reinhart, New York.

Pomerantsev, P 2017, *Nothing is true and everything is possible: adventures in modern Russia*, Faber & Faber, London.

Porter M & Kramer, M 2006, 'Strategy and society: the link between competitive advantage and corporate social responsibility', *Harvard Business Review*, vol. 84, December, pp. 78–85.

Porter, T & Ronit, K 2006, 'Self-regulation as policy process: the multiple and criss-crossing stages of private rule making', *Policy Sciences*, vol. 39, no. 1, pp. 41–72.

Posner, E 2018, *Last resort: the financial crisis and the future of bailouts*, University of Chicago Press, Chicago.

Posner, R 1974, 'Theories of economic regulation', *Bell Journal of Economics and Management Science* vol. 5, no. 2, pp. 335–58.

Pound, R 1910, 'Law in books and law in action', *American Law Review* vol. 44, no. 1, pp. 12–36.

Queensland Floods Commission of Inquiry Final Report 2014, State Government of Queensland, Brisbane.

R v *Hayes* 2015 (Sentencing remarks of Mr Justice Cooke, 3 August).

Rakoff, J 2019, 'The problematic American experience with deferred pro-secution agreements', *Law and Financial Markets Review*, vol. 13, no. 1, pp. 1–3.

2014, 'The financial crisis: why have so few high-level executives been punished,' *New York Review of Books*, 9 January.

Reinhart C & Rogoff, K 2009, *This time is different: eight centuries of financial folly*, Princeton University Press, Princeton.

Ross, E 1907, 'The Criminaloid', *The Atlantic*, January, pp. 44–50.

Rossouw, G & van Vurren, L 2003, 'Modes of managing morality: a descriptive model of modes of managing ethics', *Journal of Business Ethics*, vol. 46, no. 4, pp. 389–402.

Rousseau, JJ 2005, *The social contract*, Penguin Books, London.

Rothschild, E 1994, 'Adam Smith and the invisible hand', *The American Economic Review*, vol. 84, no. 2, pp. 319–22.

Royal Commission into Misconduct in the Banking, Superannuation and Financial Services Industry 2018, *Interim report*, Commonwealth of Australia, Canberra.

Royal Commission into Misconduct in the Banking, Superannuation and Financial Services Industry 2019, *Final Report*, February, Commonwealth of Australia, Canberra.

Ruggie, J 2010, Report of the Special Representative of the Secretary-General on the issue of human rights and transnational corporations and other business enterprises, *Guiding principles on business and human rights: implementing the United Nations 'Protect, Respect and Remedy' framework*, United Nations, New York.

Salz Review, 2013, *An independent review of Barclay's business practices*, Barclays, London.

Sandel, M 2012, *What money can't buy: the limits of markets*, Penguin, London.

Schumpeter, J 1943, *Capitalism, socialism and democracy*, Allen & Unwin. London.

Schwartz, A 2014, *Broadcast hysteria: Orson Welles' 'War of the Worlds' and the art of fake news*, Hill and Wang, New York.

Senate Permanent Subcommittee on Investigations 2013, *Offshore profit shifting and the U.S. tax code – Apple*, Library of Congress, Washington, DC, 21 May.

Serious Fraud Office v *Rolls Royce PLC and Rolls Royce Energy Systems* 2017, 17 January.

Skinner, A 1999, 'Analytical introduction' in A Smith, *The wealth of nations*, Penguin, London.

Smee, S 2018, 'Net loss: The inner life in the digital age,' *Quarterly Essay*, vol. 72, December.

Smith, A 2006, *The theory of moral sentiments*, Dover Books, Minuela.

2005, *The wealth of nations*, Penguin, London.

Snyder, T 2018, *The road to unfreedom: Russia, Europe, America*, The Bodley Head, London.

Standing, G 2011, *The precariat: the new dangerous class*, Bloomsbury, London.

Steiner, G 1971, *In Bluebeards castle: some notes towards the redefinition of culture*, Faber & Faber, London.

Stephan W & Finlay, K 1999, 'The role of empathy in improving intergroup relations', *Journal of Social Issues* vol. 55, no. 4, pp. 729–43.

Stigler, G 1971, 'The theory of economic regulation', *Bell Journal of Economics and Management Science*, vol. 2, no. 1, pp. 3–21.

Stiglitz, J 2008, Evidence to House Committee on Financial Services, 'Regulatory restructuring and the reform of the financial system,' United States Congress, Washington DC, 21 October.

2002, *Globalisation and its discontents*, W. W. Norton, New York.

Fitoussi JP & Durand, M 2018 *Beyond gap: measuring what counts for economic and social performance*, OCED Publications, Paris, 27 November.

Sen, A & Fitoussi, JP 2009, *Report by the Commission on the Measurement of Economic Performance and Social Progress*, OECD Publications, Paris, 14 November.

Stone, C 1981, 'Corporate vices and corporate virtues: do public/private distinctions matter?' *University of Pennsylvania Law Review*, vol. 130, pp. 1441–509.

Stone Sweet, A 2006 'The *new* lex mercatoria and transnational governance', *European Public Policy*, vol. 13, no. 5, pp. 627–46.

Stout, L 2007, 'The mythical benefits of shareholder value', *Virginia Law Review*, vol. 93, no. 3, pp. 789–809.

Streeck, W 2016, *How will capitalism end? essays on a failing system*, Verso, London.

Schmitter, P 1985, Community, market, state – and associations? the prospective contribution of interest governance to social order', *European Sociological Review*, vol. 1, no. 2, pp. 119–35.

Strunz, S 2012, 'Is conceptual vagueness an asset? arguments from philosophy of science applied to the concept of resilience', *Ecological Economics* vol. 76, April, pp.112–18.

Sunstein, C (ed.) 2018, *Can it happen here: authoritarianism in America*, Dey Street, New York.

Sutherland, E 1940, 'White collar criminality', *American Sociological Review*, vol. 5 no. 1, pp. 1–12.

1949, *White collar crime*, Holt, Reinhart & Winston, New York.

Sztomka, P 1999, *Trust: a sociological theory*, Cambridge University Press, Cambridge.

Talley, E & Strimling, S 2013, 'The world's most important number: how a web of skewed incentives, broken hierarchies and compliance cultures conspired to undermine LIBOR,' in J. O'Brien and G. Gilligan (eds.), *Integrity, risk and accountability in capital markets: regulating culture*, Hart Publishing, Oxford.

Taylor, C 2007, *A secular age*, Belknap, Boston.

Teubner, G 2002, 'Breaking frames: economic globalization and the emergence of lex mercatoria', *European Journal of Social Theory*, vol. 5, no. 2, pp. 199–217.

Thomson, I & Boutilier, R 2011, 'Social license to operate' in P. Darling (ed.), *SME mining engineering handbook*, Society for Mining, Metallurgy and Exploration, Littleton, CO.

Thoren, H 2014, 'Resilience as a unifying concept', *International Studies in the Philosophy of Science* vol. 28, no. 3, pp. 303–24.

Ulrich, P 2008, *Integrative economic ethics: foundations of a civilized market economy*, Cambridge University Press, New York.

United Nations, 2004, Commission on Human Rights, Sub-Commission on the Promotion and Protection of Human Rights, Working Group on Indigenous Populations, Twenty-Second Session, United Nations, New York.

Vance, JD 2016, *Hillbilly elegy: a memoir of a family and culture in crisis*, HarperCollins, New York.

van't Klooster, J & Meyer M 2015, *Ethical banking – a primer: why banks must put social purpose at the heart of their strategies*, White Paper, Centre for Compliance and Trust, Judge Business School, Cambridge University, Cambridge, October.

Vargas Llosa, M 2015, *Notes on the death of culture: essays on the death of culture and essays on spectacle and society*, Faber & Faber, London.

Varoufakis, Y 2017, *Adults in the room: my battle with the European and American deep establishment*, Vintage, London.

Walker, B, Holling, C, Carpenter, S & Kinzig, A 2004, 'Resilience, adaptability and transformability in social-ecological systems', *Ecology & Society*, vol. 9, no 2, p. 5 (online).

Wallestein, I, Collins, R, Mann, M, Drerluguian G & Calhoun G (eds.) 2016, *Does capitalism have a future?* Oxford University Press, Oxford.

Weiner, J 1964, 'The Berle-Dodd dialogue on the concept of the corporation', *Columbia Law Review*, vol. 64, no. 8, pp. 1458–67.

Welch, M 2014, 'Resilience and responsibility: governing uncertainty in a complex world', *The Geographic Journal*, vol. 180, no. 1, pp. 15–26.

Westbrook, D 2010, *Out of crisis: rethinking our financial markets*, Paradigm Publishers, Boulder.

Wilburn, K & Wilburn, R 2011, 'Achieving social licence to operate using stakeholder theory', *Journal of International Business Ethics*, vol. 4, no. 2, pp. 3–16.

Williamson, O 2000, 'The new institutional economics: taking stock, looking ahead', *Journal of Economic Literature*, vol. 38, no. 3, pp. 595–613.

Winkler, A 2018, *We the corporations: how American businesses won their civil rights*, Liveright, New York.

Wolin, S 2016, *Politics and vision: continuity and innovation in Western political thought*, Princeton University Press, Princeton.

Woodward, B 2018, *Fear: Trump in the White House*, Simon & Schuster, New York.

Yamigishi, T & Yamigishi, M 1994, 'Trust and commitment in the United States and Japan', *Motivation and Emotion* vol. 18, no. 2, pp. 129–66.

Yannis, Y 2017, *Adults in the room: my battle with Europe's deep establishment*, Bodley Head, Oxford.

Yeats, WB 1921, 'The second coming', in S. Heaney (ed.), *WB Yeats: poems selected by Seamus Heaney*, Faber & Faber, London, 2000.

2000, *Poems selected by Seamus Heaney*, Faber & Faber, London, 2000.

Zimbardo, P 2007, *The Lucifer effect: why good people do evil things*, Rider Books, London

Žižek, S 2017, *The courage of hopelessness: chronicles of a year of acting dangerously*, Allen Lane, London.

Acknowledgement

My father, an architect, gave me the capacity and courage to dream, while my son Justin has supervised a physical building on Inishbofin Island which is worthy of Heaney's ideal. To Thomas Clarke for his friendship, and my family, I cannot express enough gratitude. By their very nature, harbours are points of arrival and departure. This journey towards trust and what it means could not have been completed without their support. Thanks also to Inishbofin and its people for welcoming me home. To trust is to stop running. With this book, I have.

Cambridge Elements ☰

Corporate Governance

Thomas Clarke
UTS Business School, University of Technology, Sydney

Thomas Clarke is Professor of Corporate Governance at the UTS Business School of the University of Technology Sydney. His work focuses on the institutional diversity of corporate governance and his most recent book is *International Corporate Governance* (Second Edition 2017). He is interested in questions about the purposes of the corporation, and the convergence of the concerns of corporate governance and corporate sustainability.

About the series

The series Elements in Corporate Governance focuses on the significant emerging field of corporate governance. Authoritative, lively and compelling analyses include expert surveys of the foundations of the discipline, original insights into controversial debates, frontier developments, and masterclasses on key issues. Its areas of interest include empirical studies of corporate governance in practice, regional institutional diversity, emerging fields, key problems and core theoretical perspectives.

Cambridge Elements ≡

Corporate Governance

Elements in the series

Asian Corporate Governance: Trends and Challenges
Toru Yoshikawa

Value-Creating Boards: Challenges for Future Practice and Research
Morten Huse

Trust, Accountability and Purpose: The Regulation of Corporate Governance
Justin O'Brien

A full series listing is available at: www.cambridge.org/ECG

Printed in the United States
by Baker & Taylor Publisher Services

Printed in the United States
by Baker & Taylor Publisher Services